D1109393

TEDBooks

The Great Questions
of Tomorrow

DAVID ROTHKOPF

TED Books
Simon & Schuster
New York   London   Toronto   Sydney   New Delhi

TED Books

Simon & Schuster, Inc.
1230 Avenue of the Americas
New York, NY 10020

Copyright © 2017 by David Rothkopf

All rights reserved, including the right to reproduce this book or portions thereof in any form whatsoever. For information address Simon & Schuster Subsidiary Rights Department, 1230 Avenue of the Americas, New York, NY 10020.

TED, the TED logo, and TED Books are trademarks of TED Conferences, LLC.

First TED Books hardcover edition April 2017

TED BOOKS and colophon are registered trademarks of TED Conferences, LLC.

SIMON & SCHUSTER and colophon are registered trademarks of Simon & Schuster, Inc.

For information about special discounts for bulk purchases, please contact Simon & Schuster Special Sales at 1-866-506-1949 or business@simonandschuster.com.

For information on licensing the TED Talk that accompanies this book, or other content partnerships with TED, please contact TEDBooks@TED.com.

Interior design by: MGMT.design
Jacket design by: Chip Kidd

Manufactured in the United States of America

10 9 8 7 6 5 4 3 2 1

Library of Congress Cataloging-in-Publication Data is available.

ISBN 978-1-5011-1994-1

ISBN 978-1-5011-1995-8 (ebook)

This book is dedicated to Carla
And our never-ending conversation
Which regularly explores great questions and
Constantly reminds me
She is the answer to many of the most important ones

CONTENTS

The Great Questions
of Tomorrow

*If I had an hour to solve a problem and my life depended on it,
I would use the first fifty-five minutes determining the proper
question to ask, for once I know the proper question, I could solve
the problem in less than five minutes.*

Attributed to Albert Einstein

When I was seventeen years old, I watched a documentary
on nuclear winter that described the imagined aftermath of
World War III. Hundreds of millions were dead. Hundreds of
millions more were displaced or starving or slowly dying of
radiation poisoning.

The summer day on which I was presented with all this was
especially beautiful. Nonetheless, the film had left me in a kind
of fog, shocked by the enormity of what seemed a plausibly
imminent and horrific future for all mankind. Catastrophe
palpably loomed even for leafy, green Summit, New Jersey,
prosperous, oblivious, and bathed in sunshine as it was. I went
looking for my father.

My father was a scientist. He had come to the United States
in 1939 at the age of thirteen. He and his parents left one step
ahead of the Nazis. Almost all the rest of our immediate family,

almost three dozen people, were lost to the senseless horrors of the Holocaust. Five years after my father arrived he was in the US Army on his way back to Europe. There, he would command artillery batteries, and later embark on a journey to find the traces of lost relatives amid the desolation of the recently liberated concentration camps.

You would have thought that enduring such horror—growing up in fear in Nazi-controlled Austria, seeing his father dragged off on Kristallnacht, fighting in the bloodiest war in the history of mankind, adjusting to life after loss and dislocation in a new and unfamiliar country—would have produced a certain grimness in my father, perhaps even pessimism. It did not. That does not mean that he forgot what had happened. Every year on the anniversary of Kristallnacht he would send my brother and sister and me reminders of the day. Right now, as I write this, the last handwritten note he ever sent me—a postcard of Dachau, Germany—hangs taped to the wall over my desk. It reads "November 10."

My father channeled his energies into science. Some people found solace in religion. Others found it in expressing their feelings or sense of alienation via the arts. He chose with each blow to become more rational, to turn not to God as limned in dogma and ritual but as manifested in nature. He sought understanding in facts, proofs, and algorithms. Specifically, he sought to understand how the human brain worked, how we learned. He asked questions. Constantly. Finding the right questions and seeking answers became his mission and, I believe, his own personal religion. His work brought him to perhaps the foremost scientific and technical research institution on earth at that time, the Murray Hill, New Jersey, headquarters of Bell Telephone Laboratories.

Bell Labs was a sprawling campus. In 1973, I had a summer job there. It was a world in which scientists were given free rein to do pure research, and the results in the nearly half century since the Labs were established had already proven to be transformational. Unlike other corporate labs or sponsored research in think tanks, this was a place that, in its heyday, was about creative minds being set free to first search for the most important questions, then for the best possible answers.

Bell Labs is where radar and telecommunications satellites were developed and where new frontiers in computing were explored. Perhaps its greatest accomplishment was the development of the transistor, an innovation designed to amplify long-distance calls that went much further in its applications, ushering in the information age by enabling the miniaturization of electronics and the establishment of the networks and the creation of computing power on which modern society relies.

I'm sure that, back then, I hadn't the slightest idea of where the innovations that began at the Labs would lead, of the vast changes they would invite and demand of us all. Nor did I recognize the new questions that they would raise, or how they might tie into my contemplation of nuclear oblivion.

After searching for a bit, I found my father. If memory serves, I tracked him down at the swim club that many others from the Labs belonged to. He was by the tennis courts—a place typically dominated by a couple of fierce mathematicians with nasty groundstrokes and matching temperaments. As I approached him, he knew I was troubled by something. He was, after all, a psychologist. That said, when he asked what was wrong, I don't think he expected me to say that I was shaken by the looming prospect of global thermonuclear war. It just didn't seem like that kind of afternoon.

We sat down and I explained what I had seen. "Hundreds of millions of people will die!" I said, "And it could happen. It could happen any minute. In fact, it probably will happen." The US and Russian militaries were on a hair-trigger setting. Missiles were waiting in their silos and onboard submarines lurking just off our shores and theirs.

He paused for a minute and then asked with a kind of contrarian perversity that I know he and many other scientists thought was wryly charming, "I see. So, what is it that has you so upset?"

This kind of curveball question had been thrown at me my whole life. But still it made my head spin. "What do you mean, why am I so upset? The whole planet will be devastated. Hundreds of millions will die. Even if you survive, there will be no point in going on living."

He paused for a moment and stared out into the distance. "Well," he said calmly and with a slight trace of a Viennese accent, "you know, a hundred million people—a third of the population of Europe—died during the fourteenth century of bubonic plague. The result was the Renaissance."

The Renaissance represented a civilizational watershed. It produced changes of scope and profundity that touched every aspect of human lives. The nature of states and the rules that governed kings and kingdoms, their relationship with the Church, the fundamental tenets of religious belief, the nature of work and of economics, the nature of war and of peace, even the basic philosophies which societies embraced regarding the role of individuals, individual rights, the nature of the social contract, the very purpose of civilization would be rethought and changed forever. It was an upheaval that for Europe and, ultimately, the world, was epochal. How does the seemingly distant past, a time that for us is depicted only in the yellowing pages of books and in the cracked images of centuries-old frescoes, have any relevance in this era of virtual reality, big data, mapping the genome, and an entire world that seemingly has its eyes glued to glowing screens?

As was the case during the fourteenth century, we too are living in what might be described as the day before the Renaissance. An epochal change is coming, a transformational tsunami is on the horizon, and most of our leaders and many of us have our backs to it. We're looking in the wrong direction. Indeed, many of those in positions of power and their supporters are so actively trying to cling to the past we can almost hear their fingernails clawing at the earth as they try to avoid accepting the inevitable and momentous changes to come.

Technological shifts will be only a part of the cascading disruptions associated with the new era. As history shows, these shifts will, in turn, change human behaviors, open new areas to human understanding, enable new forms of creative expression, empower new means of economic activity, and inspire new thinking about the way lives and governments and businesses should be organized. These changes will empower the reweaving of the fabric of our lives much as the steam-powered looms of the Industrial Revolution did not only with textiles but with the lives of workers, the rise of a new middle class, the empowerment of unions, the recasting of politics, the remaking of the relationships associated with colonialism, the shifting of the power of nation-states, and so many other changes.

In every area of our lives—whether we are rich or poor, residents of a great city or a desolate region untouched by technology—it seems certain that disruptions on a similar scale are coming. Indeed, they are already beginning.

1 The Day Before the Renaissance

If we sense that such changes are coming, we have an urgent responsibility to ourselves, our families, and our communities to prepare for them. How do we begin to address these massive shifts in nearly every facet of our lives? How can we begin to prepare for changes that are of a scope and substance that may be greater than any faced for twenty generations, some that may be so great that they force us to reconsider our most fundamental ideas about ourselves and our world? And how can we shift our focus away from the old, comfortable formulations about how societies are organized and operate, what they look like, who should lead them, and what course corrections are essential?

Asking the right questions is where to begin. And, if we look to the history of epochal changes, we can see common characteristics among them that can help us understand what those questions are.

In Search of Perspective

It is an irony of life that, when our senses are most attuned to the events transpiring around us, real perspective is most elusive. Later, even as memory plays its tricks on us, nature affords us the compensatory blessing of context. While we may not remember every bit and piece of what happened, we gain a clearer perspective through the passage of time.

Imagine that you lived during the fourteenth century. It would be very hard to have much long-term perspective. During the outbreaks of the plague, survival was the only priority. And of course, as bad as it was, the plague was hardly the only concern. The Little Ice Age was beginning. The Great Schism was dividing the Catholic Church. Mongol rule was ending in the Middle East. The Hundred Years' War had begun. Dynastic upheaval in China ushered in the beginning of the Ming Dynasty. The Scots were fighting for independence (some things never change). Much as it is today, Christian Europe and the forces of Islam were in conflict, resulting in the Battle of Kosovo in 1389.

With the advantage of hindsight, we can see that every one of the shocks that rocked the era led not just to substantial progress, but to reordering of the basic way in which society was viewed. In the wake of the human losses of the Black Plague, labor became more valued, and a middle class began to emerge. Trade flows, which may well have brought the plague to Europe from Asia, also led to the exchange of new ideas and materials, to economic and intellectual growth. The nature of work and how we thought of economics began to change.

A weakened Church began to be challenged by reformers. States began to emerge in forms like those we know today. Ultimately, within a few hundred years of the change, principalities gave way to nation-states, which in turn were locked in a power struggle with the Catholic Church. The nature of governance had also been transformed.

Universities and scholarship began to take root and spread in new ways. Education for commoners, beginning with literacy, began to spread. Combining the rise of the middle class and the needs of new governments to win support from other powerful members of society, the seeds of more democratic government were planted in places like Europe, themselves predicated on a changing view of the rights of individuals and of states, the role of law, and the nature of communities.

With new technologies of navigation and new networks of roads, not only did societies interact with each other differently, thus instigating changes in cultures, but so did the nature of warfare—navies grew more important, and gunpowder and other new technologies of fighting ushered in the end of the reign of knights and local warlords. With governments and political systems changing, substantial shifts in the nature of the diplomacy required to resolve such conflicts also took place.

In other words, while the average citizen of the fourteenth century saw struggle and chaos, changes were afoot that would redefine how people thought of themselves, who they were, what a community was, and of the nature of basic rights, of governance, of work and economics, of war and peace. To understand the future and how it would be different from the past, it would therefore have been essential to consider the questions associated with such changes. To ask in the context of the changing world—and to ask again, amid what was to follow—how does all this change how I view myself, my community, my rights, my government, my job, and the way the world works around me?

During another such time of upheaval, we too would benefit from considering similar questions because, once again, the coming changes will be profound.

Of course, asking the right questions, and getting the right answers, is easier said than done. We have loads of biases. We expect the world will confirm those biases, and we mishear and misread events around us as a result. We expect the future to be like the past. (After all, we live in a world in which 85 percent of the time the weather tomorrow is the same as the weather today.) We are also harried and so busy reacting to the demands of the moment, much like the average citizen of the fourteenth century dealing with war and plagues and climatic catastrophes, that pausing to get to the root questions often seems like an impossible luxury, and one we are ill-prepared for. This last point is also important. If we don't understand the technologies or other forces at play in changing our world—be they the burgeoning sciences of the early Renaissance or the neural networks or cyberthreats of today—then how can we possibly understand what is to come?

Furthermore, if those who are supposed to lead us don't understand the changes, they can't ask the right questions either. What's more, they typically have a vested interest in resisting the questions. The status quo got them where they are, and they have a strong interest in preserving it. For example, the predisposition of our political leaders to seek to capitalize on the fears of the moment to advance their self-interested desire to cling to power regularly leads us to keep our eyes on yesterday's headlines rather than on the horizon. Fearmongering is not only exploitative—and, by the way, plays right into the hands of some, like terrorists, who seek to promote fear—it is also a potentially fatal distraction from the bigger risks associated with potential coming changes for which we are ill-prepared.

I know some of this from personal experience. In the late 1990s, I founded a company devoted to using the power of technology to help top policy makers and business leaders get the answers they needed. My proposition was that now, thanks to the Internet, we can use sophisticated tools (this was before Google) to find any answer you need. Seemed like a gold mine. It was not. Why? Because what we discovered was that the big problem in most organizations was not finding the answers, it was getting our would-be clients to figure out the right questions. I've spoken to many top intelligence officials who acknowledge that they have the same problem in the US government—vast apparatuses with huge resources devoted to gathering information, but real problems when it comes to arriving at the questions that might make those people and satellites and computers useful. One current top intelligence official said to me, "Asking the right question is the biggest challenge we face. People typically let the immediate past shape their questions—how do we avoid another shoe bomber is an example, when that's not a risk that we're likely to face. Or they let their area of expertise and their desire to be useful shift their focus. This is kind of the when-all-you-have-is-a-hammer-everything-looks-like-a-nail problem, and it leads people who feel the future is drone warfare to ask questions that end in answers that require drone warfare.

Or, to choose an example, it leads people who have spent much of their adult lives fighting Saddam Hussein to ask questions after 9/11 about his role, even though he didn't have one. And that did not turn out well."

So, in the end, Hamlet had it wrong. "To be or not to be" is not the question. The question of questions is, "What is the question?" In this respect, history tells us to start with the basics, the foundational questions that we have for too long taken for granted. There are questions like: "Who am I?" "Who rules?" "What is money?" "What is a job?" "What is peace?" and "What is war?"

One lesson is that the more profound the changes, the more basic the questions we should be asking; it is the simplest and most direct questions that cut to the fundamental issues of life, that resist nuance and evasion and rationalization more effectively.

A question like "Who am I?" can lead to questions about how we derive our identity and, in a connected world, how that and our view of communities is likely to change. The answers to those questions can lead us to question whether our old views and systems of governance for communities will work as well in the future, or whether they need to be changed. And they will also raise questions about the role of technology in helping to implement those changes, in creating other kinds of communities, good and bad, driven by our search for identity that might also impact our lives in profound ways.

While there are, as always, a few bright minds out there pioneering new ways of thinking and starting to ask the right questions, it is the responsibility of *all* of us as citizens to see the questions raised here or in similar discussions as more than intellectual exercises. Our futures depend on getting this right— as individuals, as communities, as nations, and as a civilization.

We should not be afraid of this task either, even though many of us have a natural fear of the unknown. You cannot be intellectually rigorous in your analysis of where we have been without coming to the conclusion that where we are going will be a better place. Rather than the fear that seems to have suffused so much of our discussion of the recent past, the fear that is exploited as a tool by so many leaders, it is my deeply held belief that asking the right questions about tomorrow's world and looking for those who are helping us to arrive at the answers is a certain path not only to greater understanding of the massive changes to come but also to greater optimism about the world we are bequeathing to our children and grandchildren.

In the end, at least, that is the message I think my father was trying to convey to me that summer afternoon in 1973, and it is certainly the one that he, like his innovative colleagues at Bell Labs, and a lifetime of study have left me with. Thanks to human ingenuity, most of the changes history has brought, large or small, whether they seemed catastrophic in the moment or not, have ushered in progress in its many forms and led us to the better world in which we live today—a world in which people live longer, are better educated, are healthier, have access to more opportunities, are wealthier, and have every reason to be happier than any prior generation in history.

If we can keep that in mind, then perhaps not only will we not shy away from these questions, but approach their answers with real imagination, not only wondering what might be possible, but by helping to make better outcomes more likely by considering them and becoming their champions.

2 Who Am I? Identity and Community Reimagined

One question defines us: Who am I? The answer is elusive. A name or gender or race does not provide a wholly sufficient answer, nor does a role in a family or a job description. How do we define our essential self?

From childbirth, an enormous portion of our identity is defined in relation to other people: whom we know and who knows us, the context in which we are viewed and how we wish to be viewed. We tend to view ourselves reflected through the eyes of others. We see the world largely in terms of those in our daily lives. First, the eyes are those of our immediate family. Later, we define ourselves by associating with friends and classmates. As adults, we derive our identity in a similar way, from the families we make for ourselves, our coworkers, people we associate with politically, and people in our clubs or activities. To some extent, we also derive identity from those with whom geography, culture, or religion gives us some affinity—our neighbors, our countries, those who share our traditions and beliefs.

When I identify myself, it is often in terms of my personal relationships: as the romantic partner of the great (and patient) woman with whom I live, or as a father of two wonderful, creative, often hilarious, and loving daughters, or as a son or a brother proud of the family he grew up in. Professionally, I am, or have been, a writer and a CEO and an editor and a government official. Each of these dimensions of me requires someone else, or a group of people, to round out the identification: family, colleagues, readers, partners, constituents. No one of these is sufficient to completely identify an individual; rather, each is a piece of the mosaic that, taken together, equals our points of view—our "selves"—and philosophers through the ages have suggested that other, deeper forms of reflection and peeling away earthly things, experiences, and relationships are the keys to discovering an entirely different awareness of who we are.

Throughout human history, these associations have been driven by one factor more than any other—proximity. We saw ourselves through our associations with those who were literally as well as temperamentally, emotionally, politically, or spiritually *close* to us.

Now, in the connected world and that of the approaching era of machine intelligence, that is changing.

The New Town Square
Thanks to the ubiquity of cell phones and the connectivity of global networks, effectively every person on the planet will be connected to one another in a man-made system in a few years' time. This will be a historical watershed of profound importance. It is, as we have discussed earlier, on par with the Renaissance, though I think that it is probably much larger in its significance, because it will touch so many more people, and change so many more lives much more profoundly. Imagine: the entire world will truly be a single community, one in which it will be infinitely easier than at any time before for anyone

anywhere to reach out and connect with anyone anywhere else anytime. We are reweaving the fabric of civilization.

Let us take a moment to appreciate the staggering speed and scale of this change. The printing press with moveable metallic type was introduced for the first time during 1377, in Korea, during the Goryeo Dynasty. It took another sixty-five years or so for the idea to be first presented in Europe in the form of Johannes Gutenberg's handiwork. It was not until 1800 that the invention resulted in the printing of 1 billion books. In contrast, portable cell phones went from zero to 1 billion subscribers in about twelve years. To grasp the suddenness of the onset of this era, note that it took seventy-five years to go from zero telephone users to 50 million. It took Facebook three and a half years to reach the same milestone. It took the video game Angry Birds thirty-five days to do likewise.

Half the cellular devices being sold now are smartphones, which make up about 80 percent of all Internet usage. What that means is that people are doing a lot more than making calls and playing Angry Birds on those phones: the phones are changing the way people spend and manage their money; identify, make, and maintain friendships; participate as citizens; become educated; stay healthy; keep up to date with news; and be entertained, as well as commit crimes, incite revolutions, and seek new disruptive innovations. Even the poorest continent on earth, Africa, is expected to have approximately 80 percent cell phone penetration by 2020.

We are likely to go from perhaps 20 billion devices on the Internet today to an estimated 50 billion in 2020. Most of those devices will be embedded in things—in your refrigerator, in cars, in industrial machinery, in jet engines, attached to sensors in bridges, and in buoys out at sea—and all will be capturing data and processing it in real time, giving us an extraordinary kind of heightened, real-time awareness of life and commerce on this planet that even today we can hardly imagine. Compound that with the enhanced power with which new computers will be able to assess that data, and the ability to harness thousands or millions of computers to handle massive analyses in unison, and you can see the era of machine intelligence approaching.

There have been few watersheds quite like it, and none that have hit the earth with such suddenness or caught the planet so ill-prepared for the consequences of the changes being unleashed.

The implications for our identities are profound. Soon, we won't just be connected to every person on the planet but to every business, to a huge universe of machines and sensors that can tell us how life is being lived across the globe and in real time. We will have one global cultural ecosystem for the first time in history—where everyone, everywhere can touch, influence, and, more than ever before, relate to anyone anywhere, all the time.

Because the connected world is rewiring the ties that bind us, it's going to force billions of people to reassess who they are. We will interact with new friends and partners, view distant societies, acquire new skills and abilities to take advantage of the connections that can serve or threaten us. Each of us will find ourselves closer to worlds that once seemed distant, familiar with experiences and people who once seemed foreign. Even those who choose to burrow into the Web—to associate only with the like-minded—will find that their connections will be less bound by distance or contained by national borders.

Meanwhile, the generation that has grown up from birth with these connections already takes these modes of interaction as a given. Their town square is virtual and infinite, a place where they can meet and interact with and get to know anyone, anywhere.

Just two decades ago, there were only three groups on the planet that could be measured in billions of people: the Chinese, Catholics, and Muslims. Today, they are rivaled by virtual communities: Facebook, Google, and Yahoo users. Since we derive identity from the groups we associate with, and, since such groups develop their own cultures, rules, and vernacular, it is reasonable to ask how these virtual megasocieties are impacting our own view of ourselves, and the day-to-day reality of who we are, how we act, and with whom we interact. How are our identities evolving?

It might seem easy to dismiss all this as technophilia—an overemphasis on a wave of new gadgets that'll have a secondary effect on our lives, like ATMs or dishwashers. Yet the question we need to ask is: How might technology change our own character or that of our civilization? In other words, when the old barriers fall between me and new sources of friends, partners, collaborators, and the networks of associates that help us to define ourselves, we must ask: "Who am I?"

Modern Tribes

Now, let's look closely at one of the aspects of identity on which much of history has turned—our place as citizens of a tribe, a city, a state, or a nation. Along with religious identity, this aspect of who we are has been the source of conflict and bloodshed and mayhem since the dawn of time. It has become the basis on which society is organized. The words we use to describe our compatriots—from "countryman" in English to *Landsmann* in German—literally suggest that we are whom we share geography with.

Not any longer. "Who am I?" will increasingly be answered less by answering "Who lives nearby?" or "With whom do I share local customs?" and more by "Who shares my beliefs or my likes?"; "With whom do I share psychic rather than geographical space?"

Like the town square of old, the ease of engaging with people across the globe is giving rise to virtual communities that can share interests, develop social bonds, and coordinate efforts. When proximity is no longer primary, other important barriers fall as well. On the Internet, it becomes possible to enter new communities with different standards, to redefine or mask old or traditional definitions of oneself, and to strip away the kind of inhibitions associated with face-to-face interactions. Local community standards that once might have made a relationship or an interest or a subject taboo no longer apply.

This dissociation can be both liberating and dangerous. A new name, a new persona, allows for fewer inhibitions. So, too, the Internet allows those whose views are so idiosyncratic that they are unlikely to be matched anywhere else in a local environment to find like-minded counterparts. To take just one example, the It Gets Better Project was started by syndicated columnist and author Dan Savage and his partner in 2010 to provide a support system for lesbian, gay, bisexual, and transgender youth. What started as a single YouTube video grew into a movement whereby individuals around the world could share their stories, find support in their struggles, and unite in solidarity. Imagine the challenges to convening such a group that once existed in a judgmental society and how much easier it is to do with the reach, speed, and privacy of the Internet. Instant communities, be they refugees or oppressed minorities, of the politically like-minded or of scientists trying to tackle a common problem (the original purpose of the precursors of the Internet), are now proliferating daily.

Unfortunately, this new connectivity can also produce negative consequences. You know the stories: in the shadiest corners of the Dark Net, parts of the Web that are both difficult to reach and completely anonymous, a host of criminal actors from pedophiles to traffickers to terrorists is able to communicate and plan, to connect and collaborate invisibly. Alienated, angry men and women in search of a mission and a way to express their unhappiness with the world have connected in chat rooms and via social media; the result is the recruitment boom that has fueled the rapid rise of the Islamic State.

Even in more normalized parts of society, the question remains: With ever increasing connectivity, will the real intimacy of live human interaction suffer? Will the ability of

Internetizens to conduct themselves face-to-face or to manage complex social situations deteriorate? In 2010, Japanese government data revealed that 700,000 of its citizens suffered from *hikikomori*, defined by Japan's Ministry of Health, Labor, and Welfare as "people who refuse to leave their houses and isolate themselves from society in their homes for a period exceeding six months." The Internet, it seems, offers an enabling tool to keep individuals with *hikikomori* isolated in their own world and to deepen their sense of disconnect from everyone else. Unfortunately, *hikikomori* is a critical contributor to Japan's high suicide rate among young men devoid of hope and unable to reach out for help.

It remains to be seen how deep or enduring such changes will be. Certainly, it would be a mistake to assume all the changes to come are for the better.

However, it is not the purpose of this book to advocate for sweeping civilizational changes. They are coming whether we like it or not. The issue is whether we are prepared for the changes—how they will impact the very nature of who we are. Can we take advantage of what is best about them? Can we minimize that which is negative?

A Common Language

The unifying power of the Internet can strengthen, but also destroy, what makes us unique. The desire to live together in a hyperconnected world, to communicate and work with people, may require a certain global assimilation in which some of our distinctions, for better or for ill, are subsumed into an unprecedented human universality. I'm speaking here of language.

Throughout history, when people seek to connect to advance their self-interest, they start by finding a common language and setting common expectations. Five hundred years ago, there were perhaps seven thousand languages being spoken in the world. Today that number has fallen to five thousand, representing twenty-five major linguistic groups. Of those, a few are dominant. Perhaps one in two people on the planet speaks a language characterized as belonging to the Indo-European linguistic family.

Now, the Internet is accelerating the process of linguistic consolidation. Of the top 10 million websites in the world, approximately 55 percent are in English. The next largest number are in Russian (5.9 percent) and German (5.8 percent). While the number of non-English websites has grown steadily during the past decade, English remains dominant, with one study showing that it has remained the language of approximately half of all websites since 2005. Among the top 10 million websites, only eleven languages besides English are represented by more than 1 percent of the sites.

By another metric, as of late 2013, English-speaking users made up 28.6 percent of Internet users, Chinese-speaking users made up 23.2 percent, and then, after that, the distribution falls into single-digit numbers, with Spanish at 7.9 percent and seven others with 2 percent or more.

The writing is on the screen (and, likely, in one of a handful of languages). This means that when people want to take advantage of all that connectivity—which they will need to do to survive, to do business, to participate in society, to be educated, and to get health care—they will need to do so in one of the very few languages that will be commonly spoken in connected space.

New technologies are likely to make language translation easier and perhaps slow the trend toward a lingua franca on the Internet. The Internet also has the power to preserve cultural idiosyncrasies by simply increasing access to information, materials, and people from one's own background and exposing broader audiences to them. But, as we have seen with music and movies, most content on the Web is also limited to a few languages, representing a few cultures. English dominates the global market for movies, television, and music. A study conducted by PricewaterhouseCoopers predicts that the US will dominate the market in 2019, with approximately 32 percent of all entertainment and marketing around the world. The top ten largest corporations in the field, according to *Forbes*, are incorporated in either the United States or the United Kingdom.

Questions of cultural homogenization are inflammatory worldwide, as we naturally lament the loss of rich traditions. But the impact and economic value of Web content is linked to the size of audiences, so more is produced for those who consume the most. The sheer size of the libraries of such content is an incentive for others to become fluent in those languages.

Another question is: How important is language to identity, and do we gain more than we lose by removing barriers to communication? I think that we do, and that we fetishize differences that have caused eons of pain at our own risk. Nostalgia for divisions between people is one of the most dangerous—and frequently exploited—forces in history.

A Paradoxical Tension

Language, of course, is just one of the dimensions of national identity at risk in the post-geography world. Today, Internet users can go venue shopping for cultural commonalities—be they religious, musical, artistic, or political. The paradoxical tension between the power of the connected world to help preserve or redefine the local or to create new forces of homogenization is an essential characteristic of this new era. It can facilitate the mobilization of flash mobs to protect a shared value or advance a shared cause. Alternatively, it can enable people to choose the criteria for the community they wish to occupy in the cyberlandscape. Arithmetic helps. When you have the world to choose from, you can build large communities of people with very specific interests or combinations of values.

Homogenization and localization are not even countervailing forces in this world. Consider modern, global elites—the richest, most powerful citizens of the world. Through interviewing the leaders of the world's largest companies and financial institutions (for books of mine, like *Superclass* or *Power, Inc.*), I found that these jet-setting participants in global markets often had more in common with one another than they did with the citizens of the countries in which they were born or lived. They read the same few newspapers (the *New York Times* and the *Financial Times*), websites, and books; dressed in similar ways; vacationed in similar places; stayed in the same hotels; visited one another in a very limited number of exclusive neighborhoods; and ate in a limited number of celebrated restaurants. Hop from one of those restaurants to another around the world on any given evening, and you would likely hear very similar conversations, because interests and sources of information overlap so much.

This is not just a product of the culture of being a 1 percenter. It is also a natural expression of their self-interests. Steve Schwarzman, CEO of Blackstone, the private equity group, once said to me that, thanks to the global network he had built up, he was one phone call away from reaching out to anyone in power anywhere on the planet. In other words, developing the most powerful people into a closely knit network serves the professional, economic, and political interests of each member of the elite.

Unsurprisingly, top financial professionals and other elites globalized faster than other segments of the planet's population because they had access to the technologies of connection earlier. While connections are nothing new, connections that extend everywhere, easily, are. And they do not simply manifest themselves as facilitated backroom chatter. They have, as in the case of Blackstone and private equity, led global finance to open capital flows to every corner of the world much faster than other industries have globalized. Why? Because every new market they wired in became a new opportunity to do business—visibility led to liquidity, which led to profits. And the lynchpin of visibility was connectivity. This, in turn, has driven global growth and led to the promulgation of more universal application of the rule of law (as that is required by bankers to protect their interests in those capital flows). At least, that's one argument the proponents of globalization (and I'll admit that I am one) have made. Having said that, linking markets and elites in those markets also enables them to mobilize where their interests align, and, because they are powerful, this has led to a disproportionate focus in the making of global regulations and laws and tax regimes on serving the interests of those who are most powerful and best positioned to wield global influence and on ignoring or underplaying the interests of the more disconnected, less well-informed, less globally influential average citizens.

New technologies, therefore, can and do provide special advantages for elites in the world today to connect with one another and to grow the gaps between themselves and digital have-nots—which is precisely what has happened on a planet where a few hundred of the families control more wealth than the bottom two-thirds of the planet's people. Will these technologies fail to become the tools of a new democratization? Will there emerge a new technological superclass that always maintains special advantages over the rest of the world, especially in areas where money buys tech power and expertise that gives special advantages like the access to and analysis of big data, security, or financial analytics and management? Or will growing access to the Internet and cheap but powerful tools provide greater transparency for the world at large into the ways of elites and ultimately serve as a leveler? This is as critical a set of social questions as we are likely to encounter in the era ahead.

More pertinent to those of us without private jets: Does this palpable acceleration of globalization mean that you might soon be less likely to describe yourself as American or Chinese or French? Probably not. In fact, the free flow of cultures worldwide—made possible not just by the Internet and new communication technologies but also by modern transportation—has created a backlash. The rise of Islamist extremism is largely motivated by a desire to preserve ancient traditions in the face of encroaching Western values. A flow of refugees, who have been enabled by new technologies even when the passages they make have sometimes been dangerously low tech, also produces a backlash and a rise of nationalism. Technology is obliterating borders, and some people who feel that their identity is in flux and are clinging, in vain, to keep it rigidly fixed.

It's clear that our newly connected world has the power to both homogenize and help solidify the local. The question should not be which of these forces will prevail; indeed, in my view, we should expect to live with contradictions and tensions like these in the future, just as we have in the past.

Perhaps the solution rests with the myriad individuals and organizations that are using new technologies and the opportunities they create to help ensure that a *convergence* of civilizations is the more likely long-term prospect than a clash of civilizations. We see evidence of this new reality all around us, whether it is a social media–savvy pope, whose millions of followers are gaining spiritual insights in statements of 140 characters or less, or in the breathtaking goal set by Facebook to have 5 billion members on its site by 2030, creating a community greater than any nation, one whose rules are being set by techies and businesspeople rather than elected or selected officials. These individuals are using the networks of wires linking the world and its communities as a kind of loom; each individual piece may be unique, but the result weaves together the fabric of a new global society.

I Think, Therefore I Am. Or Not?
Of all the challenges to our traditional views of our own identities that we are likely to encounter in this new era, among the greatest will be the emergence and evolution of artificial intelligence. *Evolution* is a resonant word in this regard. A tipping point looms for our very idea of what it means to be intelligent, what intellectual capacity is. The past limits we ascribed to intellect—those of the human brain—will be transcended. Intelligence beyond ours is possible. In the whole history of humanity, there has never been a moment when our species did not possess the most powerful intellectual capacity on our little planet. That is a distinction that is not likely to survive this century.

Philosopher Nick Bostrom has been at the forefront of considering the implications of this looming watershed. Based on his years of work with leaders in the field of artificial intelligence, he has become convinced that profound changes are just ahead. He asked those leaders when they thought that we would have machines that could truly think independently of the men and women who made them. Some answered 2030. Some said 2040. Some said 2050. No one said never or centuries from now. This is just a couple of ticks of the historical clock away, just a generation from now.

Bostrom poses important questions about how the emergence of this new force within our society is likely to take place. As Bostrom has posited, if we can create an intelligence as great as our own, what comes next? If that intelligence surpasses ours, what will the machines that possess the capability to reason and innovate do next? Surely they will not stop from seeking ways to create even greater machine intelligence. At what point, and in what ways, do we defer to greater machine intelligence?

Professor Sherry Turkle at MIT is another visionary who has spent much time reflecting on how we interact with our technology, how we develop strong emotional bonds with and dependencies on comparatively primitive artificial intelligence even today. Imagine, she asks, how we might feel when robots are caring for our children, replacing breadwinners, or delivering medical care. How will we conceive of ourselves when sentient robots or superpowered computers assume many more of our roles within society? Or how will we use machine intelligence and technologies to enhance our own capabilities? We already know that brains can be wired to machines (or to other brains) to enhance human performance. How would augmented human intelligence be made available? What would become of social and economic divides if it were not available equitably? (Let's be honest: it very likely won't be.)

Artificial intelligence is quickly moving beyond mere machines and is slowly but irreversibly being woven into the rest of the fabric of our lives. Consider what might happen with AI-empowered virtual reality.

Rony Abovitz has. He is the president and CEO of Magic Leap. A son of an inventor and an artist, Abovitz's virtual reality headset does not replace the world in which we live with a fantastical boardroom—rather, it overlays our reality with a digital one. "Imagine," he recently said, "you are walking in China, and all the billboards are in English. And at restaurants, as people are talking to you, there are live subtitles," he says, explaining what the device will enable. "You don't even realize you're in a computer; it's just happening." Virtual buttons and sensors could appear all over your home, letting you know when you are running out of milk or laundry detergent and giving you instant access to ordering more from a supplier like Amazon.

As virtual reality expands, it allows us to conduct more of our daily lives in this created space that is the intersection of the real world and an overlay of artificial sports arenas, schools, marketplaces, and hospitals. Think Pokémon Go writ large, a seamless interweaving of the digital and natural worlds. There may even come a time when the computers we build shape our environments without any human commands. We will certainly be required to act in our physical world, but what will it mean to live when more and more of our lives takes place in artificial space? Will it become harder to tell the real from our facsimile of it, when the bulk of our lives is conducted by clicking on overlaid computer buttons? What, then, is real? Where is home? Which world is more comfortable for us? We could ask whether humans could become more acclimated to the artificial environment, but we have already established that they can and do. After all, the artificial world has been created by men and women to embody the planet of their fantasies, the one they wish they could occupy, or at least the one that might entertain or empower them.

AI is also finding its way into our very beings. Medical researchers are developing digestible biosensors to monitor a host of bodily functions. Dr. John A. Rogers is a leader in the field of bioelectronics, which is the intersection of electronics and organic systems. With over one hundred patents to his name, Rogers recently accepted a position at Northwestern University in Chicago, where he will continue working with state-of-the-art nanotechnology to develop soft electronics with silicon semiconductors flexible enough to operate with the human body to give instant updates of medical information to patients and their medical teams.

MIT-trained futurist and inventor Ray Kurzweil projects the success of these endeavors and their introduction into our most essential human systems. He writes and speaks widely about his theory that this technology will become so sophisticated in the coming decades that our craniums will hold not only our brains but countless nanobots infused with our DNA scurrying around our capillaries. These tiny machines will enhance our own mental processes with nonbiological thinking. While others think of such a cyborg future as replacing the human with the mechanical, Kurzweil sees a brighter future for humanity: artificial intelligence will improve upon both rational and emotional thought by literally connecting our brains to the Internet. If his predictions come to pass, we'll be able to enhance our thinking and draw from wider sources when contemplating big ideas, searching for innovative solutions, or simply exploring a new genre of music.

With our experience being so thoroughly infused with artificial intelligence, what will it mean to be human, to exist? Will these tools, as Kurzweil predicts, simply build upon our humanity, improving our ability to express our personhood? Will they lead to improved efficiency in our businesses and associations by allowing meetings to take place anywhere at any time, with each participant's ideas enhanced by online resources and instant access to resources? Or will we slowly recede into artificial intelligence, the machines creating our worlds and fueling our thoughts? What is the answer to the question "Who am I?" when the thoughts framing it emanate from tiny computers created and programmed by someone else?

We can go further. What if a machine were deemed to be self-aware, as intelligent as a human, possessing a kind of consciousness or even just possessing unique knowledge of benefit to society: How might the law evolve to protect the machines and/or the knowledge within them? Can we envision a day in which smart machines have rights? And what if the machines outpace our intelligence and reasoning abilities? What would we make of humanity then? What would they make of it? Would such questions gradually become less relevant as humans gradually become seen as simply an evolutionary link to higher, machine intelligence?

Nick Bostrom has recently made a career of thinking about this future in which artificial intelligence surpasses, and possibly stands in the place of, our own. Before that day comes, he contends, we need to teach artificial intelligence to adopt our values. As machines develop and learn, we can establish initial conditions and program the initial goals to ensure that machines develop so as not to threaten the things we hold dear.

As you can see, in countless ways, the coming changes to our society will force us again and again to consider the basic question "Who am I?". But they are not just forcing such a reappraisal; they should also be enabling it. What once our neighbors or our elders or centuries of custom might have crushed, connectivity and ubiquitous communications are allowing to flourish. Multiply such changes across all the facets of each of our identities and the result is profound. Touch every person on the planet with such changes, accelerate the ways in which such changes can occur, identify all the new aspects of ourselves we will be able to examine in this world, and the promise is greater still. Of course, many will seek to use emerging technological resources to counter the trends of self-empowerment and self-awareness and growth. That struggle will certainly be a hallmark of the decades ahead, especially in societies that seek to remain closed and to fight the changes that are surely coming.

3 Who Are We? The Social Contract and Rights Reconsidered

"Who are we?" is an even bigger question than "Who am I?" We are social animals; we live our lives among one another, depend on each other, must contend with those around us. To facilitate this, when we left what philosophers like Thomas Hobbes and Jean-Jacques Rousseau described as the state of nature, our ancient ancestors entered into a social contract. With this social contract, we agreed that we were better off together, but the trade-off was that we had to play by a certain set of rules: not stealing whenever we wanted something, not killing each other whenever we felt angry. In short, the terms of the social contract demand that we subordinate our individual needs to the needs of the community. While most of us do not spend much time thinking about those terms, they establish the essential framework for our lives. To be effective, they must evolve as civilization does, adapting to new realities, acknowledging new challenges.

As we adapt our social contract, we must return to the core questions—the questions we seldom ask but ought to. While politicians might, often rhetorically, ask "What kind of society do we wish to live in?" the bigger question, the one that dates back to the very origins of the social contract, is: "Why do we live within a society at all?" It is one we have lost sight of in many ways. Indeed, acknowledging the importance of the question "Who are we?" raises related concerns: "What are our rights?"

or "How should we govern ourselves given new technologies at our disposal and the way technologies have always altered the nature of governance?"

We the People . . .
National constitutions are supposed to enshrine fundamental rights for everyone—and for generations. Such documents are also, of course, products of moments in time. That's why the best of them, like the US Constitution, draw from the debates of the era, like freedom of the press, and contain the seeds of their own reinvention. Indeed, the secret to a sustainable constitution is that it both captures what is enduring and anticipates the need to change. Among the questions most important to ask, therefore, is: "Are we and our institutions and philosophies flexible enough to adapt to change, to the impending ethical and moral and social challenges for which there are no good precedents?" I believe the answer is yes. (Though, as I consider it, I can't help but think of Winston Churchill's famous remark regarding the United States, in which he expresses confidence that we always do the right thing, although, typically, "after [we] have exhausted every other possibility.")

Over the years, the US Constitution has been amended twenty-seven times—the first ten being the Bill of Rights—to stay current with prevailing views of what is fundamental or best for the United States. Among the finest examples of the Constitution's adaptability to shifting and maturing norms are the Thirteenth Amendment, which ended slavery, and the

Fifteenth and Nineteenth Amendments, which guaranteed voting rights for everyone regardless of race or gender, respectively.

Because it is meant to be malleable, the original Constitution included references to very few technologies. In fact, America's founders were so sure that technologies would evolve over time that they even included protection for the rights of innovators in Article 1, Section 8 (the Copyright Clause). The technologies that were mentioned were ones that, by the late 1700s, had become so ingrained in daily life that they seemed natural, or at least critical, to the functioning of the government: money, for instance, and a military. In at least two cases in the Bill of Rights, the unfettered use of technologies was necessary for citizens' freedom—those technologies being the press and arms. The press was more than three centuries old when the Constitution translated into the law of the land in the United States the freedom of expression. Meanwhile, the arms referenced were not specified, but no doubt included the firearms of the day that were essential to the upkeep of a militia, which was the express rationale for the right to bear arms in the first place.

Technological progress challenges the assumptions that underlie even the best-conceived documents. This has been evident recently in the debate over whether Fourth Amendment guarantees against illegal search and seizure, which explicitly pertained to the main information storage system of the late 1700s (papers), now cover technologies like email and metadata. And, tragically, while there has been vigorous public debate noting that the arms of the twenty-first century could not possibly have been foreseen by the authors of the Constitution and have little to do with a well-regulated militia, special interests in the United States have resisted changes that might have better maintained the spirit of the narrow, purpose-driven right conferred by the Constitution in terms of the technological realities of the present day, a world effectively without militias or the need for them.

Arguing that people cannot assert rights beyond the imagination of the Constitution's framers is an absurdity as great as suggesting that every law conceived in the context of the eighteenth century should apply hundreds of years later. Further, it is dangerous. It is hazardous not to bring the American conception of rights in line with the ways and means of modern life, because to do otherwise creates vulnerabilities and risks inconsistent with the principles underlying the Constitution. (Surely the framers of the Constitution did not intend that the right to protect one's records or home from unreasonable searches only protected individuals from the kinds of searches that existed at the time the document was drawn up.)

The technological changes afoot today demand a reconsideration of what constitutes a fundamental right. What's more, we dare not let risks or threats that are so complicated or new that only a few truly understand them grow simply because they are arcane—nor is it consistent with our idea of democracy to allow only those technological elites who understand an emerging issue to define how it is resolved while leaving the ignorant masses to live with the consequences of their conclusions (as is currently happening with much of the debate around what is acceptable government surveillance behavior in the connected big-data world).

The Right to the Internet

Unfettered Web access is the modern equivalent of the right America's forefathers demanded to read or distribute a pamphlet written by Thomas Paine or to gather in a town square to discuss it. The UN Special Rapporteur for Freedom of Opinion and Expression has argued that disconnecting people from the Internet constitutes a human rights violation. Estonia passed the first law enshrining this right in 2000, but the trailblazing country's engagement with the Internet dates back a decade earlier, to the midst of a crisis. In 1992, facing a failing economy in the wake of the Soviet Union's collapse, Prime Minister Mart Laar invested in digital systems for business registration and new government infrastructure. Internet use was widespread before 2000, with every school in the country connected by 1998. Since that time, access to the Internet has spread to remote parts of the country, promoted efficient government reforms, and attracted globally minded investors. Connectivity is viewed as an essential part of the modern identity of that country, which has resulted in it being a trailblazer in considering the new issues associated with this new era. Other countries, including Costa Rica, Finland, France, Greece, and Spain, have since asserted some right of access in their constitutions or legal codes or via judicial rulings.

As connectivity expands from people to devices—the Internet of Things—all those interactions must be covered by our laws and founded in some sense of our values. Who will write these new laws? Who will debate them? Who among our electorate will be able to understand them when they are implemented?

Related questions rise into view. In the Information Age, does the right to privacy exist? In an age when the fundamental units of economic value are bits and bytes, who owns the data an individual or a sensor may produce? If there is a right to the Internet, for instance, does that mean people must also have a right to the electricity needed to plug into the Web? The answer, resoundingly, is yes. Electricity once seemed a luxury, but today, the nearly 1.3 billion people without it are effectively cut off from modern life. This raises another question: In a world where roughly 80 percent of the electricity is—and for a long time will be—produced by burning fossil fuels, how is the right to a clean, healthy environment balanced with the right to the social and economic participation afforded by the Internet?

Who Owns the Data?
Right now, profound decisions about who can tap into the Internet, how much of the Web one can access, and who lawfully owns the data being contributed, shared, and used across it are being made, often without benefit of forethought. Those decisions are being made by powerful interests who would be perfectly happy if you did not think too hard about questions like "What should the nature of privacy be in the connected world?" or "What are the rights of individuals and governments regarding their data in the connected world?" or even a fundamental question like "Who owns my data?"

Data flows are becoming as important to the competitive success of nations, companies, and individuals as capital flows. Understanding this, a foundational question must then be

"Who owns the data?" Imagine that each person emits a small cloud of data that told of their interests, financial resources, health, shopping needs, and so on. That data has significant economic and political value. It's a value that big companies have been trying to induce average citizens to give away since the dawn of the Internet Age. Google, for example, has persuaded hundreds of millions of people that offering up their data in exchange for free email makes economic sense—and then sells the data they have gathered for billions. It will go down in history as one of the great scams of all time, akin to buying the island of Manhattan from Native Americans for $24 worth of beads.

The question "Who owns the data?" is more complex, however. Some of the data we provide—about our location or political affiliations—might be of special interest to an intrusive or oppressive government. Thus, you can imagine a situation in which corporations, governments, and individuals all want to claim a right to access this data. That is a battle being waged right now: the efforts that the US government has made to try to gain access to encrypted data on an Apple iPhone as a way to fight terrorism; the efforts that companies have made to have privacy laws written so that data collected in their apps or software or hardware might be something that they could access and monetize.

If bits and bytes, therefore, are the most basic value-carrying instruments of the new economy—the coin of the realm of the Information Age—then how you answer the question "Who owns the data?" determines who has the wealth in this new era, and who has much of the power.

How Often Does the Social Contract Need Updating?
These changes seem certain to shift how society works—
redefining haves and have-nots, power and its absence and its
use, what opportunity is and how it is gained, and the nature
of risk. Knowing that such changes are afoot, we then must ask
ourselves how we go about adapting our social contract to the
needs of the future. All the metrics we use to measure the success or failure of our countries or leaders tend to imply reasons
that are tied to assumptions that have emerged from history
or have been crafted by special-interest groups to serve their
needs. For example, we sometimes measure our progress by the
economic growth of the countries in which we live—the speed
with which gross domestic product or income levels rise. Or we
focus intently on national measures of progress and dismiss
as foreign global trends or issues that once might have been
remote but today could have direct impact on our way of life.

But is the purpose of society purely economic growth, especially growth that benefits the few, as often happens in the case
of the metrics we most commonly use? Do events somehow
matter less if they occur outside our borders? At one point in history, a community was defined as those physically close to us;
those were the people with whom we had the most in common.
However, in the wired world, as we've seen, communities are
no longer principally defined by geography or proximity. The
nature of interaction in those communities changes. Common
interests are different. They may be greater than those within
cities or nation-states. What if they are? To whom or to whose
flag do we pledge allegiance then?

The evolution of our social contract does not just happen. In the past, most of the changes that altered the social contract unfolded slowly. It took three and a half centuries for the printing press to change social discourse to the degree that democratic governments were forced to enshrine freedom of the press as a basic human right. During that period, those in power sought to control the dissemination of dissenting views. Revolutions were fought. Lives were lost. Pioneer champions of free speech, whether political or religious, economic or artistic, were condemned to prison or even burned alive. Debate over their fates coursed through parliaments and universities and the salons of the day. Ultimately, philosophers grappled with the questions that were underlying this ferment—about core principles that had gone unchallenged: the divine right of kings, whether sovereignty resided in monarchs or in the people, what rights and powers those people had or should have. Some of these issues remain unresolved. Some countries still, for example, deny people basic rights that were long ago accepted as inalienable in Western societies. Nonetheless, great strides were made around the world.

In fact, gradually, as a consequence of many concurrent forces (chiefly mass literacy), the many gained the power to demand recognition of the free exchange of views and ideas. Freedom of the press was enshrined as a basic right of all men (and later of all women).

The path to conclusions about free expression—a value many citizens of democracies consider basic today—was not a straight nor an easy one. That should not be surprising, given how many nations in the world today still deny freedom of expression and are trying to use new technologies to stifle or distort it, perceiving this right as a threat to their power. Yet, those who can spread ideas and win support for them via new technologies find new power at their disposal. This relationship—between ideas and power—begs vital questions. Who writes the social contract? Who modifies it? When is it up for review? How do we implement changes? Are there new challenges the connected world presents us with regarding a society in which people live much of their lives in virtual space, beyond borders, beyond many of our existing rules, in a realm increasingly populated not just by people but also by machines and apps that act like people, where social behaviors are modified dramatically and even an absolute concept like truth has come to be viewed as relative or malleable?

It is at this fault line that so many of the battles of history have been fought. It is a place where the rights that are asserted and granted determine the fate of civilizations, where the accumulation of power determines the destiny of millions or billions of people, where the speed of change shakes the foundations of our institutions. And these are questions that once again—as they did during the last communications, transportation, and intellectual revolutions that remade our ideas of communities—must be asked. Indeed, they will be answered whether they are asked or not.

This reality creates a special responsibility for every one of us. If we were to look out the window in the morning and see storm clouds on the horizon, it would fall to us to prepare for a rainy day. The same is true now, only more so, because what we see on the horizon is not a change in the weather that will pass but a change in our way of life on the planet. We must ask ourselves what the changes will mean for us, how to adapt, how to prepare our families, and how to prepare personally and professionally.

Of course, grappling with such upheaval is not a job for everyone. There are disciplines that can help society adapt to these changes and that can help us understand what these changes are likely to mean, what their impact and implications are likely to be, and what we should do to use them to drive positive change across society.

We must ask: Where are the philosophers?

Where are the people who can help make sense of these changes at the level of what is good and what is evil, what we should aspire to, and how we should protect our core values in a new environment? Throughout history it has, after all, been the philosophers—the great thinkers who dare to question conventional wisdom—who have brought crucial debates about our social contract into the public conversation. They have framed questions about the reasons we entered society and our basic rights in ways that have brought down governments and produced ideas that have become the foundations for what we think of as modern society. The right to "life, liberty, and the pursuit of happiness" is a shining example. We need those voices now, but before we can begin to debate the issues of the day, before we can begin to adapt our social contract, we must first understand what the fundamental rights of this new era will look like.

Debating the important questions has always been thorny. Take debates about freedom of expression. John Milton, author of *Paradise Lost*, and a powerful political voice during England's Puritan era, helped shape Western views on this issue. He argued that truth was best achieved through a free and open encounter of views—conversations writ large and small. However, while he defended freedoms for Protestants, he worked to deny them to those whose views he thought dangerous, from Catholics to atheists. John Locke argued that we "commiserate our mutual ignorance and endeavor to remove it in all the gentle and fair ways of information," but also that opinions "contrary to human society or to those moral rules which are necessary to the preservation of civil society" should be legally accepted. Revolutionary collaborators—and rivals—Alexander Hamilton and Thomas Jefferson also offered opposing views. Hamilton asserted that freedom of the press was "impracticable," while Jefferson wrote, "Our liberty depends on the freedom of the press, and that cannot be limited without being lost."

This is as it always has been. It is, in fact, the nature of most critical debates, and we should remember that as we discuss how the new era of the connected world will cause us to rethink and revise the social contract. The push and pull is the way we kick the tires on big changes and roll ideas around in our collective imaginations before embracing them.

No Time to Think?
Today, changes happen quickly and then rapidly spread. The pace and direction of change is fueled by a set of forces I suggested in

Columbia University's *Journal of International Affairs* in 1998, forces that affect all aspects of life in the Information Age.

These included:

- Acceleration: Events would unfold faster.
- Amplification: The impact of those events would be magnified, would touch more lives.
- Volatility: Acceleration plus amplification produces bigger swings in outcomes.
- Decentralization: Networks redistribute power (although see discussion later in this book about the network paradox, and how networks both disseminate and concentrate power simultaneously).
- Disintermediation: Middlemen are undercut and often disappear.
- Interconnection: The most obvious characteristic of networks is that they link all their members, but the implications of such ubiquitous linkage are hard to understand and take many forms.

The trouble is that, on this rapidly shifting ground, we are literally acting before we have the time to think. We are enshrining into law or promulgating as regulation views that have profound impacts on the nature of our societies without benefit of public debate or, perhaps more troublingly, of the intervention of philosophical reflection. Of course, there are certainly philosophers in the world today who weigh new theories of social contract and political science (we have already encountered one of them—Nick Bostrom—on issues associated with artificial intelligence). However, the reality is that during the past several decades the study of how technology impacts our existing social philosophies, and how philosophy should impact our approach to new technologies, has grown. (We should have the humility to recognize that this study is not new. Democritus and Aristotle developed theories about how artifacts created by humans imitate nature, and how they differ from it.)

Philosophy, of course, is more than abstract thought—it becomes the foundation for our system of laws and governance. Thus, on the legal front, pioneering thinkers around the world have found that they must ask what a modern constitution needs to contain to protect its citizens. Exploring whether our right to the Internet should be enshrined in law is one such area for consideration. Considering what system of law governs the supranational space of the infosphere when the power of governments stops at their own borders is another.

One of the thorniest and most prominent of such questions has arisen in the area of privacy. A generation is rising that is so used to posting all aspects of their lives on social media, of having everything about everyone be searchable, that many studies show millennials and younger generations largely have no expectation of privacy, or that their expectations are quite different from those of prior generations. Indeed, this a generation for whom transparency is the norm to an unprecedented degree—a fact that may well impact everything from the future of surveillance by the state or by corporations to the nature of what is fair game for journalists to publish to the question of the value of classified government information systems.

(I have had some amusing experiences on this front. Once, my open-source intelligence company presented a briefing to some admirals at the headquarters of PACOM, the world's largest military command. We displayed a picture of a North Korean missile launcher. An admiral immediately stood up and shouted, "Take that image down! This is an unclassified briefing." We calmly, and somewhat bemusedly, noted that the picture was available on the Internet. He nonetheless insisted we take it down. Later, I asked General Tony Zinni, former commander of US Central Command (CENTCOM) about this, and he noted that he once did a study that discovered that of all the classified materials he received, 80 percent were available via open sources, and of the remaining 20 percent, 80 percent were discoverable via open sources. Think what you may of the stubbornness of bureaucrats, but the handwriting is on the wall. The way governments treat classified material is going to change, and the cost savings and benefits to better policy making and mission execution will be enormous.)

A crucial dimension of the privacy discussion relates to the collection of metadata.

Edward Snowden's famous revelation about the US National Security Agency program led to two cases brought on behalf of the public against the NSA. Both cases were ruled upon in December 2013, with divergent rulings. Judge Richard Leon of the US District Court for the District of Columbia ruled in *Klayman v. Obama* that such searches are probably illegal. While challenging what he called an "almost Orwellian" practice, he allowed a stay for the government to mount an appeal. Less than two weeks later, Judge William H. Pauley III of the US District Court for the Southern District of New York came to the opposite conclusion in *ACLU (American Civil Liberties Union) v. Clapper.* Citing his belief in the necessity of collecting metadata for national security, Judge Pauley dismissed the case and pronounced the program lawful under both the Fourth Amendment and the controversial section 215 of the Patriot Act. These types of contradictions underscore the complexity around the privacy issues of our day.

These contradictory rulings also, fittingly, reveal a paradox. Progress often requires us to reassess the rules of society when those rules were written for different times and realities (consider the Second Amendment, written at a time when militias were common and assault rifles were not). However, at the same time, we can often find within old texts enduring principles; thus, if the Fourth Amendment prohibited the search of papers and records, and metadata contains much of what was once contained within such papers and records, then metadata should be protected.

Moving forward into an era of ubiquitous sensing—of everything from your clothes to the bridge you drive over to your office chair teeming with sensors producing data about your life and passing it along to countless intermediaries, all with claims of ownership or rights to it—suggests the complexity of such debates will grow greater with each passing year.

To take this another step further, some philosophers are taking on the risks and conundrums raised by machine intelligence within the laws and bounds of our social contract. Bostrom, for example, has joined with other notables, like physicist Stephen Hawking and entrepreneur Elon Musk, to enumerate the dangers associated with artificial intelligence. All were signatories to a letter prepared by the Future of Life Institute that laid out their concerns but also stated that they believed that "research on how to make AI systems robust and beneficial is both important and timely," and that there are concrete research directions that can be pursued today.

What of other questions, such as who owns the intellectual property created by machine intelligence? When are machines entitled to certain rights? What if a machine spies on someone and retains that information—is that the same as when a person does it? Even if the machine never shares it with a person? What if a machine conducts an attack? What if machines could influence the outcome of an election—acting independently?

The list of thinkers doing important work around the globe on these issues is extensive, but, except for the Snowden-NSA scandal, these ideas rarely meet the public discourse. Those debates, in fact, tend to focus on the issues of the past, wrapped in formulations of the past, because that is what politicians believe works for them. It is also because that is what politicians understand. Few are trained to understand the implications of the coming technology-driven changes. Few have the vocabulary to even discuss those issues. Therefore, despite the work of innovators who recognize that massive issues are looming, those in power who are supposed to help prepare society for dealing with them are not doing so. They aren't asking the right questions. Typically, they don't even know what the questions are.

The resultant gap between the pace of change altering the nature of society and the creation and oversight of rules for society should be of great concern. When such gaps have existed in the past, as when the Industrial Revolution changed the social order more rapidly than the elites in charge cared to acknowledge, the result was spasm after spasm of revolution stretching from the late eighteenth century through the early twentieth century. We can ill afford such upheavals. But the question is—looking at the Arab Spring or Net activism movements from Occupy Wall Street in the US to upheavals in China—whether they have already begun.

4 Who Rules? Democracy and Government Reimagined

Most political candidates run on platforms promising change. And most of the time electorates are disappointed by what happens next.

That's because governments are what economists would call "lagging indicators." In other words, they typically change only after everything around them has moved on to what's next. This could be because governments are designed to be deliberative and slow to react. It could be because entrenched special interests seek to cling to power by maintaining the status quo. Or it could be because real creativity has typically come from other quarters of society—from science and the arts, places where the status quo is seen as a force to be overcome and even rejected. All three of these things could, in fact, be part of the same phenomenon. But that does not mean that change cannot come to governments, nor does it mean that major changes are not needed now.

For all their generally glacial embrace of modernity and general lack of creative thinking, states and their bureaucracies are the primary actors changing the nature of governance and representation in the world. As philosophers and actors push the conversation about the nature of the social contract and the roles and responsibilities of government, these sleepy institutions must take their first lumbering steps to bring new ideas to fruition. Most countries, as you might imagine, are moving too slowly toward progress in the digital age.

Because our views on identity, the nature of society, and the nature of fundamental rights are changing, our view of the institutions upon which we rely needs to adapt, too. These institutions need to change in many ways, from how they view and measure their effectiveness to their structure. Profound questions also need to be asked. Do we still need government? Should governments still be defined by geography, as they were before? Should government services remain as they were before, or, ideally, can new technologies shift roles between public- and private-sector institutions and change for the better how those institutions operate? Clearly, it's time to go back to the drawing board with our most basic ideas about governments and governing.

Casting a Vote for a New Kind of Democracy
Let's take something basic: voting via the Internet. What could be a more natural idea? It would make participating in democracy easier for many citizens. It would likely increase participation rates in elections and make elections more truly representative of the will of the public. It could be faster; tabulation could essentially be constant and in real time. For those worried about hacking or the security of their vote, the response should be straightforward: It's a manageable issue. If you trust your bank account and your pension and your life savings, your most personal information and all the work you do professionally to the Internet, surely you can trust your vote.

Giant financial institutions daily conduct trillions in transactions on the Internet, and those are secure enough to protect the totality of the global economy. Our nuclear secrets and the vital national security secrets of virtually every country in the world are in large part trafficked on networks accessible to the Internet, and we find satisfactory security for them. And, to complete the argument, as we have seen, existing systems for collecting votes have had their profound flaws—from stuffing ballot boxes to hanging chads to just plain old rigging the results.

Perhaps it is time, therefore, that we took people out of the process of counting the votes altogether and left it to secure, transparent computer systems.

That said, according to the National Democratic Institute (NDI):

> It is fair to say that Internet voting is not a commonly used means of voting. Of the fourteen countries that have so far used it in any form, only ten currently have expressed any intention of using it in the future. . . . Internet voting seems to fit, for many countries, a niche corner of the electoral process.

Only four countries have used Internet voting through several election cycles: Canada, Estonia, France, and Switzerland. Again, trailblazing Estonia stands out as being, in the words of the NDI assessment of the subject, "the only country to offer Internet voting to the entire electorate. The remaining ten countries have either just adopted it, are currently piloting Internet voting, have piloted it and not pursued its further use, or have discontinued its use."

Consider the alternatives: voting machines that are kept under lock and key by the parties in power, or paper ballots counted by hand in front of observers selected by the parties—or hybrid, hodgepodge systems where different standards and techniques are used by different agencies across a single country.

The failure to more fully embrace the Internet in the voting process results in the most basic mechanism of democracy being grossly out of step with the other changes that are rocking it to its core. Candidates campaign on social media, blogs and Web-based commentary fuel public debate, and live Internet-based polling or quasi-referenda via Twitter polls or other means shape views of who is up and who is down. Grassroots organizations have been empowered because the means of distributing views are so inexpensive. Modern presidential campaigns use data analysis to identify and target the people most likely to vote or to be swing voters, and, ultimately, to conserve resources. We use the same means to judge politicians while they are in office, and to shape the public debate around the great issues of the day (and the minor ones that gain public traction). Every aspect of democracy has been changed by the advent of new technology, except perhaps the most fundamental one: how citizens can directly influence the operations of our governments through the selection of our representatives. Voters could have much more power much more often. Technology could be used to reduce the costs of campaigns and help get big money out of politics. It could be used to make the operations of government more transparent. In short, it could help us produce a better answer to the question "Who rules us?" than we have today, whereby in many lands it is those who have co-opted the levers of power and used the opacity and antiquated structures of existing systems to maintain their hold on power.

At the same time, the recent US election and elections across Europe have forced us to ask other questions about what an updated system might look like. Does making democracy truly digital create new risks? Is it more open to hacking and manipulation? Does easy voting via smartphone raise the possibility of shifting more to direct democracy where all big questions are decided by e-referenda? Does this undercut the benefits of more deliberate representative democracy? Does it raise the likelihood that electoral decisions will be more driven by emotion than thought? (Or by thoughts clouded by news consumed via social media echo chambers and "fake news," untruths accepted by large groups of people for dubious reasons?) It is almost certain that these risks are growing and must be assessed, and that we must resist the temptation to follow in every direction technology might offer to take us, especially when it strips us of checks and balances and filters that have proven to be beneficial throughout history.

Disintermediating Government

The story of governments' lagging on technology is all the more striking when you consider the degree to which technological change could transform the nature of governance. Technology can improve and simplify governance when properly harnessed. By streamlining and clarifying its functions, it can help it to better provide services to its people, dispose of unnecessary steps and secret costs, and even root out corruption. It has done away with intermediary industries like travel agencies, changed the way people shop for real estate and cars, and put an end to the brief life of video-rental outfits like Blockbuster. It can do the same thing for government bureaucracy and the headaches and obstacles it presents to basic citizen access to services. You can see signs of what is possible in different corners of the world already, experiments that hint at greater things to come.

You may personally be familiar with other benefits of e-governance from your daily life, such as the ability to pay traffic fines or municipal fees online or to get information about Social Security without having to deal with the bureaucracy directly. Singapore offers an eCitizen Portal to a whole range of government services via the Net, including applications for paid maternity leave and passports, online bill-paying tools, a directory of all government agencies, and a portal for collecting public feedback. South Korea ranked highest for e-government services for the third time in 2014. Its advanced telecommunications enable the government to develop strong online and mobile public interfaces, such as a Home Tax Service that offers 24/7 online access to a range of tax-payment and assessment-related services. India, like many other nations, has begun to lay the groundwork for offering a broader array of such services via its National e-Governance Plan.

In 2014, the United Nations ranked the top innovators in e-governance. Most of the leading countries are from the highly developed world. South Korea leads the list, followed by Australia, Singapore, France, and the Netherlands. The United

States is seventh on the list with the United Kingdom eighth. Little Estonia now ranks fifteenth, while China does not crack the top twenty-five. Countries with poor infrastructure have lagged behind, but there have been signs of progress. Six countries in Africa, where "progress ... remains relatively slow and uneven," six countries—including Tunisia, Mauritius, Egypt, Seychelles, Morocco, and South Africa—have above-average ratings.

None of this may sound revolutionary; portals are old-think on the Internet, and doing business online is commonplace. But virtually every sector impacted by the information revolution has seen one powerful, transformational trend shift expectations, and thus it's likely that the impending changes in governments will be profound.

Take the big bureaucracies of today. They are incredibly costly, making many of them ideal targets for disintermediation. It is not just payment processors at the Department of Motor Vehicles, either. Do governments need large embassy staffs when so many communications no longer pass person to person, and indeed, often bypass diplomats altogether (whose role, after all, was fundamentally that of go-betweens carrying outmoded forms of communication)? Governments once collected the best data on national economies. In the big-data era, that is a function that can much more easily be managed via machines and algorithms than rooms full of economists.

Government, in short, is poised to transform the way it interacts with citizens. It must. Which leads us to critical questions: Who should enact this change? Will changing metrics change the priorities of governments?

An Innovative New Way to Help the Poor: Give Them Money! These questions extend beyond borders, not only within them. For instance, will the way richer countries help poor countries—like development programs—change? It might well be possible to rewire how national wealth is distributed to those in need of assistance while simultaneously cutting back on bureaucracy and making the entire process more transparent and significantly cheaper. There is a vast developmental bureaucracy in the world that works to guide the flow of funds and the implementation of programs intended to help the poor. That bureaucracy, for all its good intentions, is costly, and in key cases does not work as well as a direct and simple approach. As illustrated in a wonderful article in *Foreign Policy* by Rosa Brooks, there may be a better way:

> One recent study looked at poor rural Kenyan households that were randomly selected to receive unconditional cash transfers from GiveDirectly, a US-based NGO. (The recipients were given amounts varying from the equivalent of at least two months' worth of average household consumption expenditures to about three times that.) Over a period of two years, it found that those households that received the unconditional cash transfers increased both consumption and savings (in the form of durable goods purchases and investments in their self-employment activities). They increased food expenditures close to proportionally to overall nondurable expenditure . . . and health and education expen-

ditures more than proportionally. Alcohol and tobacco
expenditures did not increase.

Unsurprisingly, recipients of the cash transfers also reported
higher rates of psychological well-being, and, when tested, had
lower levels of stress hormones.

Similar programs from Uganda to South Africa have shown
equally promising results. Think of that: more benefits, fewer
intermediaries. This change might also mean less corruption, as
it becomes possible to cut out the local government bureaucracies that often siphon away substantial amounts of aid dollars.

The same logic can apply in domestic US government programs. Indeed, it is in such programs that an even more disruptive idea lies. For example, two economists at MIT, Andrew
McAffee and Erik Brynjolfsson, suggest that, in our future
hyperproductive, machine-aided economies, there may be less
and less work for people to do. If there is less work, there is a
possibility that workweeks will shorten (much as they did from
seven days to five over the past century). The question then
becomes: How do those who work less make enough to survive?
Companies will make the same amount and may make greater
profits. In other words, those who possess technology and capital will gain greater and greater power and wealth, and those
who have historically depended on working for other people's
enterprises will have less work to do and presumably will make
less. It may already be that we are seeing signs of this phenomenon within the US and other big economies, like China, as
growth keeps happening and corporate profits keep growing,

but job growth is sluggish and wages have stagnated. (In the United States, this is compounded and reflected by the large growth in the numbers of those who have left the workforce altogether, making our relatively low so-called unemployment numbers questionable.) The traditional relationship between overall economic growth and the creation of new, growing, better opportunities for workers has broken down.

The consequence is that the rich grow richer while everyone else falls farther and farther behind. It's hard to imagine, but without new measures, the gross inequality we see today (90 percent of the benefits of the most recent recovery went to the top 10 percent of the population) is likely to grow worse. It's not far off from what largely discredited father of socialism Karl Marx predicted—but just because it came from Marx doesn't mean it's not true.

One possible answer to this problem, which also echoes Marx and therefore will trigger fierce resistance in places like the US, might be new redistributive mechanisms of taxation and social programs. Of course, such programs are disparaged for their inefficiencies and perceptions that they are unfair. However, what if new technologies enabled us to cut back on bureaucracy and have transparent, algorithm-driven systems for redistributing tax revenue not from company to government to citizen but direct from a machine-driven real-time tax collection mechanism to the end user? What if the algorithm could change the payments based on multiple variables associated with the performance of those paying in the taxes and those receiving them? New means of revenue generation are also on the horizon, as ubiquitous sensors make it possible to charge tiny, incremental taxes and usage fees in a fairer, more transparent, more progressive way. Oregon, for example, has experimented with charging driver usage fees for highways based on tracking GPS signals that let them know who was using the highways and who was not. With sensing systems everywhere in the big-data world to come, it will be much easier to create finely tuned tax policies that collect revenues in new ways—charging polluters with fees that ramp up based on how long norms were violated or by how much, for example, or making tax rates more progressive by allowing for more incremental fluctuation based on shifting economic circumstances within a year or from week to week, or targeting payments from the state to the needy in more effective ways.

Death and taxes may be the great unchanging certainties of life, but there is no doubt that the modalities of assessing taxes, collecting them, and distributing them will change dramatically in the years ahead.

The Dinosaurs in Charge

As we've seen, among the big winners in hyperproductive economies are the biggest companies.

In my book *Power, Inc.*, I explored why and how it had evolved that the world's one thousand largest companies have greater economic resources at their disposal, and likely greater global influence, than more than half the countries in the world. One reason is that giant global companies can adapt better to change than political entities can. Governments are tied down to the land beneath their feet like Gulliver in Lilliput, while, to paraphrase author Tom Friedman, huge companies float above the countries that were once their domiciles. New technologies are, as we have discussed earlier, making it possible for communities to lift free of broken political systems. While governments are tied down, everything that impacts them—from the way economies operate to the way people interact—changes. Thus, big global corporations have been able to grow to the point at which the biggest companies are both more powerful and have greater international resources and influence than the vast majority of governments. And this is a disparity that will only grow as companies are empowered by technological changes, while many governments resist or lag behind them.

We must then ask: In the world of the future, where will the real power lie? Is the era of the predominance of the nation-state drawing to a close? What will replace it? Is it in the interests of the people of the world to let profit-motivated entities without any social mandate or obligation to the public interest gain the upper hand over those institutions that are created—in theory, at least—to look after the interests of society at large? (That's a rhetorical question, hopefully.)

Wrong Leaders, Wrong Questions

Answering those questions would be easier if the people in power understood the changes that are afoot and had the context, insight, and vocabulary to frame or answer the questions they need to be asking. For the most part, they do not.

In part, this is because precious few countries have elected or appointed computer scientists or technologists or people with any sort of meaningful training in these areas to top positions: the people who understand that there are big questions out there that need answering.

Let's take the US Congress, given that it is the top legislative body in the world's most powerful country. Only 12 percent of Congress's members have a background in science or technology, according to a 2011 study by the Employment Policies Institute. And, based on my conversations with tech executives who regularly interact with Congress, just a handful of people on Capitol Hill truly understand the implications of big data, cyber, and other technological revolutions.

It's true that often, there are small pockets of designated whiz kids who are cited to show that the government is up to date (as in the Office of Science and Technology Policy or the recently created Chief Technology Officer's office in the White House). But these are enclaves that deal with a few narrow issues and have neither the resources, the political profile, nor the inclination to work with every part of government to ask the question that is regularly asked in every business and most private homes around the world: What could we do differently or better thanks to new technologies?

Turn the subject to how other next-generation scientific developments, such as those in the areas of neuroscience and biotech, will raise critical questions about how we deal with mental health, crime, extended life expectancy, bioethics, and health-care costs, and the number of well-versed legislators falls even farther. "In many cases to zero," a professor at one of America's leading schools of public health said to me in early

2014. Think about that: the very underpinnings of government are dealing with issues like public safety and public health. And we have already seen what happens when we don't understand or anticipate changes properly.

In the US, for example, we have a retirement health-care system based on life expectancies from the 1920s but involving cost structures (especially the heightened expenses associated with modern life-extending techniques in the very last months of our lives), and, as a result, our system is bankrupt and offers the lowest care of service of any developed country in the world in the vast majority of areas. Knowledge of medical science and a little related foresight could have avoided the systemic bankruptcy and failure of service we face today. However, bioscience is going to change treatments and extend lives further, and big data is going to change the way we monitor patients and provide care in profound ways. The most important step we can take to reduce health-care costs is to ensure that the elderly take their medications; in the past this was impossible to do, but with sensors today, it can be done remotely. Further, what if neuroscience could enable us to diagnose and treat behaviors that might lead to criminal activity or mental-health care costs, as it almost certainly will? Without a change in the knowledge level of our legislators and executive-branch officials on these issues, our laws and regulations and treatment programs will be obsolete. Indeed, many are.

To confirm this grim view, consider just this one example: at roughly the time this book was being written, Representative Lamar Smith (R-TX), chairman of the House Committee on

Science, Space, and Technology, spent most of 2016 challenging a National Oceanic and Atmospheric Administration study on climate change, claiming that those involved somehow doctored their results to confirm global warming. This is a level of reasoning and a disrespect for science worthy of those inquisitors who challenged Galileo, although lacking perhaps their sophistication and seriousness. It was echoed on the campaign trail by the Republican candidate for president, Donald Trump, and the view that climate change might be a hoax. Although ridiculous, dangerous, and demonstrably wrong, this view was popular enough that, after he won, President-elect Trump picked an environmental team guided by a nonsensical view of the environment and committed to doing it further damage.

Besides its lack of depth, the thinking among Washington's political power brokers also suffers from a lack of breadth. In researching my book *National Insecurity*, I looked at ten of the most prominent think tanks in Washington over a decade. These organizations produced almost twelve thousand events, papers, and research reports over that time. Of these, the majority concentrated on just a few topics—such as the Middle East, the War on Terror, and China—linked closely to whatever was in the headlines at the time. Other issues, deserving of focus but outside the buzz zone, got much less attention. The issues that got by far the least coverage? Science and technology. Never mind that they are responsible for the life-redefining changes ahead or many of the emerging threats with which humanity is grappling. The problem is then compounded by a system that discourages innovators from presenting big new ideas that confront traditional wisdom, because the resulting controversy could make it impossible for the so-called big thinkers to later be confirmed for top government jobs.

So, in Washington, DC, and in many other capitals around the world, you have the wrong people dealing with the wrong problems at the wrong time in a system that is designed to discourage them from doing any different.

Will change come to governments? Of course. But, as we have seen in the past, when governments failed to anticipate or respond to social upheavals, the changes that came were disruptive and often came at a high cost. Think of the revolutions that came during the Enlightenment (from America to France) or during 1848 or the rise of industrialization (and of reactions to it like communism). Can we avoid such turmoil? Will some states resist the inevitable transformation (like autocracies or theocracies that fear the new technologies)? For them, the question is not if they are made obsolete but, rather, what replaces them, and when. Who will lead such changes if government officials themselves do not and cannot?

The changes could bring great efficiencies, savings, and effectiveness if implemented properly, but they could also introduce new inequities or empower technological or other elites in dangerous ways. Is a new generation of revolutionaries on the horizon? Did we see the first stirrings during the tech-fueled flash protests of the Arab Spring? Or are they more subtle, working in places like Silicon Valley, where first they will remake society by changing the way we relate to one another, ourselves, our jobs, money, education, health, war, and peace, and then simply leave it to government officials to realize that they are relics of another era, their powers and prerogatives increasingly altered by events over which they had little say?

5 What Is Money? Economics, Work, and Markets Reimagined

One of 2015's biggest blockbuster movies was *Jurassic World*. But the dinosaurs on the screen, realistic as they may have been, are not the ones that should scare us the most—those are the ones in our midst, the ones who wreak havoc through their failure to evolve and adapt to change, their failure to ask the right questions. Politicians, as we have established, largely fall into this category, as do some of the lemmings at think tanks. But there is another group out there with similarly scary traits. We call them *economists*.

The term *economist* may evoke visions of kindly, bespectacled wonks droning on about arcane theories, or perhaps government big shots mumbling unintelligibly before Congress. But we know better. These are powerful women and men. They have made giant policy decisions that have affected the lives of billions, often while working behind closed doors with data and strategies that few understand and fewer still believe in.

Economics has long been known as the dismal science. Thomas Malthus, a cleric who also wrote about economics, has become the poster child for illustrating the rationale behind this label. (Thomas Carlyle coined the term in reference to the study of slavery.) In the very last years of the eighteenth century, Malthus posited that population growth would ultimately derail human society's efforts to perfect itself. "The power of population," he wrote, "is indefinitely greater than the power in the earth to produce subsistence for man." It is indeed a grim prognosis, but it highlights another reason economics might be seen as dismal: that is, just how off the mark its predictions can be.

Being wrong has long been a special curse of economists. You might not think this would be the case in a so-called science. But, of course, all sciences struggle in those early years before scientists have enough data has been collected to support theories that can reflect and predict what happens in nature. Scientists from Galileo to Einstein offered great theories, but, due to the limits of their age, labored under gross misconceptions about reality. And, in economics, we are hardly in the era of Galileo quite yet. It is like we are somewhere in the Middle Ages, where, based on some careful observation of the universe and an inadequate view of the scope and nature of that universe, we have produced proto-science, also known today as crackpottery. (See long-standing views that the Earth was the center of the solar system or the belief that bleeding patients would cure them by ridding them of their so-called bad humors.)

Modern economic approaches, theories, and techniques, the ones that policymakers fret over and to which newspapers devote barrels of ink, will someday be regarded as similarly primitive. For example, economic policy makers regularly use gross estimates of national and international economic performances—largely aggregated measures based on data and models that are somewhere between profoundly flawed and insanely wrong—to assess a society's economic health (before determining whether to bleed the body politic by reducing the money supply or to warm it up by pumping new money into its system). Between these steps and regulating just how much the government spends and takes in taxes, we have just run through most of the commonly utilized and discussed economic policy tools—the big blunt instruments of macroeconomics.

(I remember that, when I was in government, those of us who dealt with trade policy or commercial issues were seen as pipsqueaks in the economic scheme of things by all the macro-sauruses beneath whose feet the earth trembled, whose pro-nouncements echoed within the canyons of financial capitals, and who felt that everything we and anyone else did was playing at the margins.)

Think of the data on which these important decisions were based. GDP, as it is calculated today, has roughly the same rela-tionship to the size of the economy as estimates of the number of angels that can dance on the head of a pin have to the size of the heavens. It misses vast amounts of economic activity and counts things as value creation that aren't that at all. Even Simon Kuznets, who pioneered the idea of GDP in the 1930s, warned against using it as the prime measure of national eco-nomic well-being. Trade data, such as that used in measuring national surpluses and deficits, misses a big chunk of trade in services and much Internet activity, among many other swaths of trade, and is widely reported inaccurately. Labor statistics, such as unemployment rates, are cooked and deceptive. The list goes on. The reality is that only two things are dependably known about most of the data that policy makers use to make decisions: it is late, and it is wrong.

Michael Green of the Social Progress Imperative is trying to look beyond this insufficient construct of national success. "Robust economic growth," Green argues, "does not automatically translate into well-being among people." His Social Progress Index is an index based on social and environmental metrics to provide a more holistic assessment of national welfare and to provide policy makers with more insights as to how they can elevate citizens' quality of life.

Today, the world stands at the dawn of a new era, thanks to the advent of big data and enhanced computing power. Already there exist data flows that will show economic fluctuations in real time and down to an incredible level of detail: by community, by block, by family, by business, by however you want to slice it. Using these tools and new sources of data to come, the world will be able to find correlations never imagined. Old ideas, like tracking national economic performance based on geography, will give way to new ones, like tracking customizable groups that share much closer correlations than borders. There is a "You-istan" out there full of millions of people who act more like you, who respond to stimuli more like you do, and who rise and fall more like you than do your neighbors. Next-generation economists using such tools will be able to target their predictions and prescriptions more surgically.

So, as we have asked "Where are the philosophers?" we might ask "Who will be the new economists?" What will they study, and how will that study differ from that of their predecessors?

Whereas today's economic models rely on a relative handful of variables, future models will utilize a limitless number, creating opportunities for new approaches, theories, and methods. Many of these new models and tools will require not microeconomists but nanoeconomists, super-specialists in the relationships between much smaller economic units and the larger economy. Economic policy making will therefore devolve from central governments to state and local governments, which are not only closer to the issues and the solutions that workers, companies, investors, and citizens require but are also better equipped to work with the local private sector in real time to solve those issues.

New economic theories will also emerge based on growing sources of real-time data. Soon, money as we know it will be replaced by bit-based and mobile-payment systems, knocking old-school monetary policies for a loop. Your grandchildren and mine will likely never hold currency in their hands, nor will they ever go to a bank. Come to think of it, you probably very seldom go to a bank today. Those grandchildren will no doubt titter as old timers like me croak out stories about how in the days before cash machines ("What's cash, Grandpa?"), if we didn't get to the bank by three o'clock on Friday ("What's a bank, Grandpa?"), we would be broke for the weekend ("What's a weekend, Grandpa?").

A Bank in the Palm of Your Hand

The early glimpses of our financial transformation come from some unexpected places. Many emerging economies are blazing trails in mobile money precisely because they are home to the majority of the 2.5 billion adults who lack access to a formal bank account, which is almost half of all adults worldwide. Not having access to a financial institution is a serious impediment to growth. Often the poorest in the world must go great distances to cash a check (even the government-assistance checks they desperately need), they can't borrow (even though microlending programs repeatedly show they are among the most reliable borrowers in the world in terms of paying back what they owe), and they can't send or receive money outside their small communities.

A solution has emerged that is now making waves globally. In Kenya and Tanzania, millions now use their cell phones as banks. The success of such programs suggests that hundreds of millions worldwide are likely never to see a bank in their lives as more and more people switch to completely virtual banking.

This has become possible because mobile data traffic, even in some of the most challenged economies in Africa, is growing at over 100 percent per year. This was seen by big telecom companies (the banks of the future?) as an opportunity to close the financial-services gap in these markets. Vodafone stepped in with a platform, M-Pesa, that it launched in Kenya and Tanzania in 2007. From those two countries, the technology quickly spread. Customers are now using M-Pesa to send, receive, and borrow money via their cell phones in markets elsewhere in Africa, Asia, and Eastern Europe, with more expansion to come. Thus, the mobile-money phenomenon has gone well beyond proof of concept and into practice. As quoted by Jason Kohn in an entry on Cisco's *Connected Life Exchange* blog, Rene Meza, a managing director of the program, said: "The banking infrastructure in Africa is not as developed as it is in the West.

Furthermore, a lot of our customers cannot meet the minimum banking requirements. This called for an alternative transaction model. The convenience of mobile services, the adoption of mobile telephony in Africa, and the merging of technologies has created an infrastructure that allows for money transfer services—in our case, M-Pesa."

The rapidly burgeoning system offers many advantages. With people handling less cash, the risk of robberies and other crimes goes down. It becomes much easier to accept government social-service payments, and easier for the governments to get the payments to citizens. A man who effectively played the role of central bank governor in Peru described for me how it once cost more in postage to get benefits to citizens high up in the Andes than some of the checks they were sending were worth. What's more, shopkeepers where the checks were cashed knew when the checks were received, and raised prices on the days they were most likely to be cashed. The new system saved costs and stopped the gouging, what the official described to me as "removing the artificial tax on the poorest that the old system had helped create." It enables people to conduct transactions more easily and to have a more transparent view into their financial health.

One senior World Bank official told me, "We expect this will be the next major area in which we'll see leapfrogging, as we did with cell phones. Communities that had no hardwired telephone infrastructure found themselves embracing the new wireless options more rapidly because there were no alternatives to impede adoption. We can expect the same here. In fact, I wouldn't be surprised if this is one area in which circumstances will put the emerging economies ahead of developed ones in the eventual move toward a virtual banking world—one in which trips to brick-and-mortar financial institutions are increasingly rare."

In fact, such developments are happening so fast that they are blurring and altering the business models we have grown accustomed to in unexpected ways. Not too long ago, I sat with the president of the international division of the world's leading logistics and shipping company as he explained why, since they covered the entire supply chain for many customers—from product fabrication to final sale—and did so via sophisticated, digitally driven services, there was no reason that they couldn't become a bank themselves, offering financial services along the way to their clients. This is not an entirely new concept. Car companies and big-equipment providers offered leasing and financing services years ago. What is new is that it is becoming ever easier to blend the services they offer—often companies seeking to take advantage of captive customer bases with whom they work daily. Through such creativity, new possibilities are being created that will forever alter the way we see money and finance and the institutions that have, until now, defined them. In this new world, banks are invisible and held in your hand, and every company is a financial institution. It is a world without currency, and thus without central banks, and yet one in which the authors of a few algorithms that drive markets and exchange rates will have disproportionate power. In this landscape, microtransactions might take place constantly and invisibly and make megafortunes for new generations of innovators. Thus, it is also a world in which the nature and shape of crime may change, and with it our ideas of security. New laws and regulations will be needed. We'll need new legislators who understand the current issues—issues that will only come at us more rapidly in the decades ahead.

Consider companies as they enter the big-data era—for example, General Electric. It produces jet engines, power plants, and MRI machines (among other things). Each of these will be fitted with sensors that will provide real-time data on the performance of the machines, but also on their location and usage, and thereby on critical aspects of key industries and the global economy. That data can be monetized: sold not only to traditional GE customers but to others who could use it to manage their own businesses, fuel their own models, and monitor global economic performance. That data is an important asset, with a real value to the company and its shareholders. Yet that data asset does not appear on the balance sheet at GE or most other companies. Similarly, the data liabilities associated with the risks that data producers and managers encounter are also absent from the balance sheets. (Data liabilities, for example, might include potential losses that a company might suffer because of data breaches.) This is a big gap. In the big-data era, virtually every company will have a big-data asset. That means that, until accountants and managers learn how to quantify and communicate data assets effectively, every company will be undervalued or improperly valued.

Cyber-Protectionism vs. Internet Internationalists

The world's intellectual center of gravity tends to shift with its economic center of gravity—as it did from Europe to the US, and as it is now certainly doing from the US across the Pacific to Asia. That suggests that the creative leadership of the next generation is also more likely to come from Asia. That, in turn, means that the ideas that may be embraced may well be more driven by Asian values, and by the fact that the rising powers of Asia, like China, share much more with the rising powers of the rest of the world than they do with the 4 percent of the population that live in the narcissistic US. We have already seen this shift take root in a crucial debate about the Internet economy being an open or closed network.

Right now, there is an intense ideological battle between countries that believe the Internet should be open and those that believe it should be closed. Who wins this ideological battle will determine the distribution of wealth and power in the world of the twenty-first century, because it is increasingly clear that the Internet has become the primary mechanism by which global economic activity will be conducted in the decades ahead. And negotiations regarding that openness are certain to replace negotiations about trade in goods as centerpieces of international economic diplomacy.

Americans have long expected that the rules of the Internet will largely resemble the vision for the Net promulgated by its creators in the West. US companies and political leaders have promoted an idea that might be characterized as cyber-internationalism, the idea that the Net will transcend borders and drive the opening of all societies. However, an alternative narrative is emerging. The Chinese have embraced a view that might be characterized as cyber-nationalism or cyber-protectionism, which suggests that they may exercise the same controls over the Internet use of their citizens as they would any other activity within their borders. This has included everything from the creation of the so-called Great Firewall of China, which blocks sites and exchanges they deem unsuitable, to other forms of censorship and regulatory control. That may seem like a losing proposition, echoing the failures of the closed societies of the Cold War era. However, it would be a mistake to think that American or Western views will necessarily dominate. History is against such conclusions. How will these new powers shape the future of the Internet and the economy within it? Because, rest assured, it will ultimately be they who do so.

China's views do not make it an outlier; Singapore, India, Saudi Arabia, Russia, Venezuela, and Brazil have all embraced stringent controls on the Internet or the way business is conducted there: who can or cannot buy and sell products and services associated with the Web. It's clear that there is a growing cyber-nationalist movement around the world. More than twenty-two nations have added significant new controls of this type in the past two years. This movement will be central to defining the economic landscape of the twenty-first century, determining how easily Internet-delivered services or capital can flow from one part of the world to another or how challenging such future commerce will be. New rules will need to be written, and new international regimes and pacts will likely emerge.

If, as we have asserted, a central economic question of this new era is "Who owns the data?" then the mechanisms by which that ownership is determined become vitally important. In the European view, where citizens must opt in to relationships with companies that harvest their data (in other words, give their consent to let the data be taken), the default protects the citizens' right to their data. In the US view, where the model is to opt out, it means the default makes it easier for companies to reap the economic benefits of the private customer data they harvest. How that tug-of-war plays out will be another giant factor in selecting the winners and losers of the economy of the century ahead.

Global regulation of the Internet, of financial transactions in cyberspace, and of increasingly global monetary policies shaped by the shifting modalities of e-money, as well as how we regulate and value work in a world in which human labor will be seen in a much different light, are all issues that will similarly have to be addressed, questions that must be answered, before the shape of this new era will be clear.

Economics Is Being Reinvented: Time to Reinvent Economists? The questions economists ask, and the way they keep score, have an enormous impact on how we judge our leaders and our own progress as a society. We stand at the threshold of an enormous leap forward, but as we make it, additional questions loom. How should the disappearance of money as we know it alter the shape of governments and policies? How will the identification of strong correlations across borders change the imperatives for government policy makers? Who will benefit disproportionately in this new system? Can we create new ways to ensure opportunity and equity? Will instant information flows force us to be too reactive and not strategic enough? Will we be able to predict the economic future more accurately? Will that information be used for the many or the few?

Again, we can't answer the questions if they are not even being asked, if our economists are unfamiliar with the underlying forces that are transforming our economies, if our political leaders know no better, and if the business innovators who do understand the change are not seen as vital partners of their counterparts in the public sector (or, as is so often the case, they distrust the public sector and would simply prefer to be left alone). The result is a knowledge gap that leads to an enormous vision deficit. Ultimately, the crashing sound you hear is the result of obsolete systems running headlong into cold, hard realities that could have—and should have—been anticipated.

Indeed, tomorrow's economics will be so unlike that of today that it might just take a Hollywood device—like a mosquito preserved in amber, carrying the blood of former US Fed Chair Alan Greenspan, whose viable DNA can re-create this macrosaurus—for future generations to fully grasp the Jurassic Period of economic thinking and approaches that have governed and guided our daily lives. More important, as that era closes, we must recognize that another is beginning. As transitions between historical epochs in the past have shown, those who evolve are the ones who will survive. The beginning of evolution is recognizing what is changing. But the secret is recognizing *how* we must change, and then implementing those changes effectively. For companies, for governments, for economists, for investors, and for each of us who works for a living, we must recognize that much of what we take for granted will soon be eclipsed: jobs, money, economic theories, the way governments interact with the economy, how societies grow, and how they take care of their people. Getting them wrong or leaving it to the few to capitalize on the changes to the disadvantage of everyone else invites the kind of upheaval and crises that have accompanied many of the biggest transitions in history. Getting these changes right promises less burdensome, less risky, more prosperous lives.

6 What Is War? What Is Peace? Power, Conflict, and Stability Reimagined

The contrast between war and peace has ever been among the starkest known to humankind. War was violent, bloody, and left visible scars on society; peace, a time of quiet safety. Today, new technology has ushered in an era of a new kind of conflict. It's a war that, on its surface, looks like peace; it can be waged constantly, and it will be invisible to most in society. This kind of conflict might be called a cool war, one that is so inexpensive, because of its invisibility and deniability, that it may be that some powers find they can never stop fighting it. In consequence, even seemingly basic constants like war and peace should be approached with an open mind, and with the kind of foundational questions with which we must challenge and examine every new development of this new era. This, like the other examples cited, echoes watersheds like the onset of the Renaissance. In the fourteenth century, it might have seemed impossible to imagine an end to the feudal system, knights, and the horse, the culture known as chivalry. However, by 1477, the rise of the use of pikes and infantry proved devastating, leading to a decisive defeat for knights fighting in the old style in the Battle of Nancy. At the same time, newly emergent states were creating professional armies, which were cheaper to maintain and easier to equip with new technologies.

As we have discussed, change moves more rapidly in the current era. We can already see signs of the modern-day equivalents of the Battle of Nancy, the first skirmishes in cyberwar, in warfare depending on unmanned smart weapons, like drones, and other hints of what may follow. These changes raise the prospect of a new era in which warfare is constant, in which the line between war and peace is forever blurred or obliterated. In this era of cyberwar, nations may seek not to destroy one another but to weaken institutions, to wound or impede enemies without killing them. It will be invisible, but it may also elevate tensions between nations in such a way that the invisibility itself will be dangerously deceptive, raising the prospect of more traditional conflict even as it seems to sidestep it or obviate the need for it.

The Shifting Terrain of Power

In this new era, not only will the Internet be used as the delivery system for cyberattacks, it will be essential to all other aspects of warfare, from intelligence gathering to recruiting fighters to spreading propaganda that might influence the outcome of elections or tilt public opinion to assessing damage. At the same time, it will be the ultimate target, the vital organ that, if shut down, can paralyze or destroy an economy or a political enemy. Thus, the terrain of the network—defined by infrastructure, regulations, and concentration of resources and capabilities—becomes as critical as geographic terrain. In politics, security, and business, understanding that terrain and the new rules of power that pertain to it will become especially important. So, too, will understanding how elements of the network impacting its speed, ease of use, and security influence the conditions for success or failure of those on the network.

Historically, power was achieved by ruling the waves, as Britain once did, or, as more modern conflicts have required, through air superiority. In this new era, those with the greatest understanding of networks and cyberwarfare, the greatest resources to command this new terrain, will have the power. At the same time, those who are most dependent on the network—likely also to be the most developed countries—will become more vulnerable due to this very dependency.

We might call this the *network paradox*, a phenomenon by which joining a network can both strengthen and create new vulnerabilities. A corollary to the paradox, however, might be called the *network power paradox*, in which the network both empowers all those on it and enables a constant shifting of that power, creating more independence and capability than ever before, both for those at the fringes of the network (at the bottom in the traditional power distribution) and for those at the center or top in traditional hierarchies of power. In such a connected world, the flow of influence courses in every direction. The opportunities, along with the security threats, abound for us all. The very nature of power, the very nature of conflict itself, are more broadly distributed and dangerous.

A compelling example of just how subtle this new technological warfare can be is the United States' efforts to challenge oppressive and extremist narratives. In 2009, at the height of Iran's Green Revolution, Jared Cohen, a member of former Secretary Condoleezza Rice's policy-planning staff at the State Department, advocated for an aggressive social media campaign supporting the message of pro-reform groups and counteracting the repressive regime in Tehran. As the regime sought to isolate those who challenged the rigged elections, Cohen helped devise a plan to assist protesters in Tehran to remain connected to the international media. Reaching out to his connections at Twitter, Cohen asked them to delay a scheduled site upgrade to keep the protesters plugged in. Twitter complied. He almost lost his job for breaking with protocol, but Secretary Clinton jumped to his defense. (But think about how quickly the use of Twitter became broadly operationalized by all who wanted to influence public opinion and shape the fate of political movements.)

Cohen left the State Department within the year, taking posts at both the Council on Foreign Relations and then with the newly founded Google Ideas, now called Jigsaw. In an article in *Foreign Affairs* and later one he coauthored with former US Deputy Secretary of State William Burns in *Foreign Policy*, Cohen offered strategic advice to governments to challenge ISIS's online machinations and the likely proliferation of such techniques across militant groups in this era, as well as on the overall need to develop digital diplomacy as a core strategic capability of the US and other governments. An array of tools lay at the disposal of world governments, including targeted counternarrative campaigns, akin to antibullying ones. He promotes research into software to identify and weed out digital terror propagandists and empowering law enforcement and online forum moderators to do the same. These proposals challenge the sensitive divide between security and freedom on the Internet, but offer a glimpse into how Internet users and governments can collaborate on issues such as e-counterterrorism.

Given the shifting equation of power, the previous example illustrates that big companies, too, can gain power in ways that give them extraordinary influence. Consider the companies with billions of users or those that control information or key technological monopolies or near monopolies but have greater resources and global reach than many countries. Who can influence more people: a major power, like the United Kingdom, or a company like Facebook? The power that used to control the waves, or the ones that dominate the airwaves and the mind share? With an economy in which the building blocks of economic value are bits and bytes rather than acres of fertile land or vaults full of gold, whoever is best able to monetize data, make connections, capture intelligence, or create new forms of value wins and gains resources and strength. Indeed, the winners are more likely to be those who can gain the virtual high ground in multiple ways. (In 2016, the media company I run honored Eric Schmidt, the executive chairman of Google, for Google's impact on the global landscape by naming the company "Diplomat of the Year"—because we came to the conclusion that tech companies were doing much more to impact international relations than most if not all traditional national states.)

We have also seen how relatively small groups, including terrorist networks or private hackers, can wield power that once only countries had. They can disrupt or shut down economies, muster armies, steal vast sums, or wage massive and effective propaganda wars.

Who holds power is shifting. The nature of power is shifting. How power is used is shifting. When Harvard's Joseph Nye wrote his famous book *Soft Power* in 1990, the Internet as we know it had yet to be launched. Cell phones were the size of bricks. Technology was available only to the few. The changes that have taken place in the quarter century or so have redefined what soft power is. It is no longer just the alternative to sticks and economic carrots, an auxiliary concept to the practice of geopolitics. It has become the primary mechanism of power, the mobilizer of masses and of those who wage asymmetric conflict, and the shaper of belief systems. As we consider the advent of big data and artificial intelligence, it is easy to imagine a world in which he who wields the best algorithm will be able to defeat whoever fields the biggest army, a world that my colleague Rosa Brooks perfectly encapsulates in the title of her excellent book *How Everything Became War and the Military Became Everything.*

Yet again, this is a domain in which the understanding of changes and their implications is often very great both within and between governments and between governments and the private sector. An area of such importance requires that we rethink how we choose people for top government jobs, and how we shape alliances and weigh adversaries. But, of course, all our decisions must begin with seeking to understand the profound nature of what is already afoot.

Waging War from a Distance

The twenty-first century will be the era of network warfare and, therefore, increasingly the era of automated warfare, multiplying the power of the best-equipped armies in unfathomable ways. Sound far-fetched? This is currently a central focus of many of the military's best minds. A former vice chairman of the Joint Chiefs of Staff, for example, recently described for me a program he had been involved with at Harvard that considered the decision-theory issues behind swarms. Specifically, his program looked at how large groups of connected, perhaps autonomous or semi-autonomous drones might be sent to handle critical missions. These fleets would communicate among one another to reassess conditions, reassign targets when members of their group are hit or malfunction, and work together with less and less human management required.

If tech superpowers, without so much as a single human on the ground, could have the ability to deploy swarms of smart drones, smart machines, or robot waves of cyberattacks, controlling all this from far beyond the reach of defensive conventional weapons and ground-bound armies that could easily devastate these poor societies with little relative risk to themselves, we must ask: What is to stop actors across the Internet from using these tools of digital and physical destruction unceasingly? Just how out of control can this become, and what can be done to limit the negative consequences for safe usage of tomorrow's technologies for social, economic, and political purposes?

The new reality is that threats will come at all of us faster than before, from all precincts of the network. Naturally, forces with greater resources will have greater power to dominate, and those on the fringes will primarily have greater power to disrupt. The Net enables power to swiftly transfer from node to node or ad hoc alliances to emerge quickly as actors seek strength through collaboration.

In a 1998 article, I addressed the contradictory phenomena of the new era:

> The revolution . . . breaks down hierarchies and creates new power structures. It amplifies the capacity to analyze, reduces reaction times . . . and can be a tool for amplifying either emotions or rationality. . . . It can make the United States so strong militarily that no one dares fight her in ways in which she is prepared to fight, while enabling opponents to take advantage of new options in asymmetrical conflict. It cedes some state authority to markets, to transnational entities, and to nonstate actors, and, as a consequence, produces political forces calling for the strengthening of the state. It is the best tool for democrats and the best weapon for demagogues.

The article was published six months before Google was founded, six years before Facebook, and nine years before the iPhone—too early to assert what certainties the then-budding revolution would bring. Now, nearly two decades later, there is one clear certainty: contradictions are themselves an essential aspect of this new era, and should inform us as we seek to command a virtual landscape, one that we have made but whose form keeps shifting and whose horizons we cannot see.

The Cool War

We are now in the midst of what we might call the Cool War. Why cool? It's warmer than the Cold War, because it involves near-constant offensive measures that, while falling short of actual warfare, regularly seek to damage or weaken rivals or penetrate defenses. In the Cool War, these offensive measures are primarily enacted through Web-based warfare.

The speed with which change has come in the field of Web-based warfare is remarkable. In 2007, there was not a single page in the official threat assessment of the Directorate of National Intelligence addressing cyberthreats. By 2011, a report from the US Office of the National Counterintelligence Executive (ONCIX) indicated that certain countries (they cited traditional US rivals, like Russia and China) posed "significant and growing threats" to America's security and economic vitality through their cyberintrusions. That report asserted that those two powers "will almost certainly continue to deploy significant resources and a wide array of tactics" to try to level the playing field between themselves and the United States. At roughly the same time, the world was coming to know more of US capabilities and activities in this area. In 2010, we first learned of Iranian centrifuges being infected with a virus that seemingly had been introduced from abroad. By 2011, US officials like White House arms-control expert Gary Samore were giving winking acknowledgment that the US might be behind the spread of that virus. In 2012, the *New York Times* revealed that the virus was related to the Stuxnet worm as part of a joint US-Israeli intelligence operation called "Operation Olympic Games." In June 2013, Edward Snowden, a former contractor for Booz Allen working on US intelligence community projects, began to reveal documents that laid out the full scope of what the US National Security Agency was doing to spy on both US citizens and hundreds of millions of people around the world. After Stuxnet and Snowden, it was clear that we were not in Kansas anymore. We had entered a new world defined and threatened by new kinds of digital-era conflict and intelligence activities.

As David Sanger, a pioneering *New York Times* journalist covering these issues, said to me, "The day after Stuxnet was like the day after Hiroshima. We had the technology, and no one else did. But within a matter of a few years, that had changed." So had the nature of warfare and of peacetime—and, by extension, of modern diplomacy.

The Cool War has been as active and constant as any of the other, more public areas of digital disruption about which we make movies and out of whose protagonists we create heroes. It is just more shadowy. Cool War heroes look like the members of the Chinese People's Liberation Army unit 61398, the Shanghai-based operation revealed to be behind many cyberattacks on US companies and government agencies. One such attack evolved into a major intrusion into the US Office of Personnel Management, which stole data on more than 23 million government employees.

Via these attacks, the Chinese gain access to valuable insights into how the government works—or into the intellectual property of US companies that helps their own companies become more competitive. This puts them in a better position to conduct more damaging attacks—akin to the Stuxnet intrusion—should they someday wish to.

We do it to them, and they do it to us. Countries and nonstate actors and individual hackers everywhere are getting into the act. It may seem less dangerous than out-and-out military confrontation, because the goal is not to destroy the enemy but to spy on them and, in certain instances, degrade them. These new technologies make it possible to reset the risk profile of conflict, making it seemingly safer and thus more tempting for tech superpowers. Of course, such attacks raise the tensions between societies and keep them constantly at odds. When you drop a bomb on a country, it not only devastates its target but it also disintegrates. However, when you launch a worm against a facility, that worm or its elements remain intact and discoverable, and thus reusable by the victim of the attack. In other words, while cyberconflict may avoid hot exchanges, it has produced almost constant escalation.

While it all may seem more benign than the era in which my generation and those before it grew up, the era in which global thermonuclear war was a threat all the time, it is not. Because being engaged in constant conflict raises the possibility of escalation and errors, and those nuclear weapons have not gone away. The potential for devastating conflict between major powers is greater than it has ever been.

Contemplating this new form of conflict, one cannot help but be struck that, as in the nuclear age, technology has been advancing faster than the public's understanding of its implications. Do we understand the rules of the new warfare? When we can strike back with force? Do we understand how to negotiate the end to conflicts that neither side is willing to acknowledge are taking place?

Cybervandalism and the Deterrence and Doctrine Deficits

Cyberintrusions will become ever more effective and difficult to defend against in the world of big data and the Internet of Things that we are entering. With the combination of ubiquitous sensors and data-gathering mechanisms, unlimited memory, and massive processing capabilities, the planet's ocean of data is growing ever larger. Much of that data is owned or controlled by the private sector. This adds a wrinkle, given that only through public-private cooperation and a kind of openness and exchange that has yet to evolve can the data assets of a nation be protected.

One striking example of the new challenges we face is the famed North Korean attack on the Sony Corporation. The attack was apparently intended to dissuade Sony from releasing the movie *The Interview*, an American comedy that made fun of North Korea's leader, Kim Jong-un. While the attack was sophisticated and, in some ways, unprecedented—leaving Sony computers in smoldering ruins and seeking to directly influence the public discourse in the United States—what may have been most notable about it was the response offered by the United States. President Barack Obama, the first US president who has had to deal with cyber issues as a top national-security priority, was flummoxed as to how to respond. It was difficult to prove, at least in ways that could be made public, that North Korea was behind the attack. Further, it was unclear what the United States should do in response.

Obama chose, in characteristically cautious fashion, to describe what had happened as "cybervandalism." The reason for the careful word selection was clear. If it were called an attack, it would demand a commensurate response with an attack by one country on another (even though this was likely an attack from a country on a corporation). But we are living in an age in which the rules and doctrines about such conflicts are woefully underdeveloped at best. It is unclear whether a nation that is the victim of a cyberattack has (or should have) the right to respond with traditional use of force. It is unclear what standards of attribution might be required to justify such actions. It is even unclear what nations may wish to characterize as attacks.

For example, in the wake of China's hack of the Office of Personnel Management, revealed in 2015, US Director of National Intelligence James Clapper asserted before Congress that he did not wish to characterize what had happened as an attack. Why? Because he wanted to make it clear that such trolling for data was what governments could and should be able to do in this new era.

Clapper was taking advantage of the fact that we are in a

definitional phase. Everything is new. The rules are unwritten. The terminology is ill-defined. There are no doctrines. The downside of all of this is a deterrence deficit; countries are left vulnerable to future attacks because it is unclear what the penalties for those attacks will be. This, in turn, creates greater danger of future attacks. At the same time, it creates for those in the defense and intelligence communities an opportunity to help shape views on what is acceptable or not.

Naturally, their goal is to define parameters as broadly as possible, to give themselves plenty of room to work. But is that in our interest? How does it affect international relations in the long term? Does it make the world safer or more dangerous? How does it impact the role governments play in our lives? How do we reconcile differing views on what is acceptable in different parts of the world? What kinds of advantages will those who choose not to play by the rules enjoy? Do we need new globally agreed-upon standards? New rules of warfare? The digital version of the Geneva Convention? Do we need new institutions to manage global disputes in this area, much as the International Atomic Energy Agency is supposed to do with nuclear proliferation?

The Web Is Mightier Than the Sword

True transparency on the battlefield is another major challenge for the future of war. Effective war reporting helps keep the public aware of current battlefield realities—realities that will be evolving and which the public at large must contend.

Despite President Obama's vow in May 2013 to increase

transparency on light-footprint drone campaigns, the only thing that we can be sure of is that the number of innocents killed is at least one. The results are confusion and an incomplete public debate on the nature of these wars. Media and research institutions on the left and right have published incongruous figures of drone-strike casualties in countries like Pakistan and Yemen. Leaks and reports often relate incomplete knowledge and selective details about the various operations under way around the world. Reporting based on unsubstantiated comments may infect the national psyche with nefarious, sensationalist narratives.

While the new era appears to be ushered in on government opacity and public confusion, the revolutionary technological tools of our time are already offering examples for lifting the heavy curtains of secrecy and enabling the wider conversation we need to confront the truth of war in the twenty-first century.

They also are providing tools for insurgents to remain connected and to leverage their power. A good example can be found amid the rubble and despair of the crisis in Syria, and is that of a group of heroic journalists and others fighting the Islamic State called Raqqa Is Being Slaughtered Silently (RBSS). The underground group is dedicated to reporting on the cruelties of living under ISIS rule in its Raqqa stronghold at any price. The future RBSS members looked on with horror as the Islamic State imposed its order on the city, forcing women under the veil, targeting children for brainwashing and recruitment, and crucifying and otherwise executing their enemies.

Since then, RBSS members have sent photos, videos, and

reports to their contacts on the outside, reminding the world that underneath the ISIS threat are imperiled innocents. The group has provided credible information to the outside world, including accounts of the human impact of American and Russian bombing campaigns, which it believes are only after-thoughts, acceptable collateral in the greater war strategy. It released reports on a captured Jordanian pilot a month before ISIS released its infamous video of him being incinerated while trapped in a cage, and revealed a failed US hostage-relief effort weeks before the White House or Defense Department acknowledged it.

As you might expect, telling the world about daily life under the rule of a terrorist state comes with great risks. RBSS has made it onto ISIS's radar, and now faces the threat of grotesque violence for their bravery. Many have been arrested and exe-cuted, their killings recorded on social media to send a chilling message to those who would shine the light of truth on ISIS's practices. In response, some RBSS members have fled, and are supporting the group's mission from the relative safety of Western cities. They continue their dangerous work to this day, a model for resistance under repression, and for digital reporting. They have tens of thousands of followers on Twitter, and their posted videos on YouTube have been seen by hundreds of thou-sands worldwide. They may be under siege and at risk, but they are being heard and mobilizing global opposition to their brutal enemies.

I met with one of their leaders in Washington last year, when the company I run, Foreign Policy, was awarding those leaders one of our Leading Global Thinker Awards. I will never forget the sense of sadness and loss in his eyes. He looked as though he might be forty, though he was twenty-four, gaunt, with a dark beard. He spoke of death as though it were a close acquaintance. It had already claimed his brother, and it was that loss that gave him the spirit to put his own life at risk. He did not expect to live to an old age. He said he had known love once, but did not expect to again. My partner, Carla, was so moved by his story that she gave him the bracelet she was wearing. He immediately put it on, touching it gently. He gazed around the garish Washington ballroom and wondered aloud whether he would ever go back to Syria again. But when the topic turned to the mission of RBSS, whatever sense of loss or wistfulness one could sense immediately vanished. There was clear resolve in his voice.

There was, in the broader context, something else. There have always been brave voices like his and those of his comrades. But they had a new weapon that made them visible enough, powerful enough, influential enough that their impact was being felt in faraway capitals like Washington. It was technology. Courage plus technology could bear witness to atrocities and mobilize opposition as never before. Indeed, numerous organizations, including the Holocaust Museum in Washington and the United Nations, have begun to see new technologies as opportunities to stem the tide of genocide and abuse in ways that have never been possible—because we now live in a world in which the vast majority of citizens carry a television studio in their pockets, a means of sending live images everywhere, so that never again will distance and horror serve as a shield to protect the depraved.

Lifting the Fog of War from Tomorrow's Battlefields
Because the stakes involved in war are so high, disproportionate resources are devoted to developing new technologies of destruction. The argument is repeatedly made that only through greater destructive capacity can we be made safer; what we have learned is that the world always becomes more dangerous. Wars may be fewer and farther between today than at any time in the past, and while that is a development to be celebrated, we have also seen the proliferation of the most dangerous and destructive technologies and new approaches to warfare that could raise tensions and the risk of unintended military consequences.

Is the world of cyberwarfare safer or more dangerous than those we have seen in the past? Will swarms of drones and robot armies save the lives of soldiers, or only save the fighting-age people of rich countries while increasing the risk of conflicts that cost lives in poorer regions?

In this high-stakes area, again, our best defense against the worst possible outcomes is asking the right questions before it's too late. These questions must be posed by our governments and militaries, updating the law and international norms to provide safeguards against new forms of total war. They must be asked by our journalists, reporters, and experts, probing into the realities of digital and high-tech warfare for the facts and dynamics worthy of greater public conversation—and fighting and resisting the efforts of some to use information warfare to carry the fog of war into daily life through the spread of disinformation and the devaluation of truth. And, finally, we in the public at large must recognize and take advantage of the fact that the same tools that make this new warfare possible and empower nonstate actors like terror groups can also empower each of us and all of us to raise questions, spread our views, and create political movements to pressure leaders into thinking about what kind of future we collectively seek, what kind of peace we want our children and grandchildren to enjoy, before embracing all the new and complex tools for the wars of the future.

What is the purpose of society? How do we define who we are? What is the role of nations? What is fair? What is the role of private actors? What is the future of global governance?

That such great questions are posed by the cascading consequences of the digital disruptions already taking place is a sign that we are indeed at an epochal watershed. Today is, in fact, the day before the Renaissance. While the changes ahead are exciting, they also demand that we accept an urgent responsibility: long before we can gain any real answers or insights into what that future might be, we first have the obligation to undertake a harder, more consequential search—a search for the right questions.

It is a prospect that is as daunting and, at times, truly terrifying as it is promising. But that is the nature of great threshold moments in history. Fortunately, we can take comfort from the fact that any rigorous analysis of that history demands optimism. Progress works. Today we live longer, healthier lives, and more people are better off, better educated, and more empowered than ever before. There is greater freedom and greater hope for the future.

The problem is this: Progress has as often been a great shock to our system as it has been pure inspiration. And it is compounded by the fact that great strides forward—from the advent of the Age of Exploration to the Industrial Revolution—have also brought with them exploitation, upheaval, and sometimes tragedy, from the slave trade to colonialism to genocide, from the abuse of workers to the failures of communism.

In the end, the difference is made by those who rise up in each era and lead with ideas. For the world, the challenge is to find them and sort through them to separate those with something beneficial to offer from those who are leading us in a more dangerous direction.

Perhaps somewhere out there is the next Marx, coming up with an alternative to modern capitalism based on the growing inequity he or she sees within it. He or she won't look a thing like the last one, nor will his or her ideas necessarily resemble those of Marx. He or she probably won't be found as Marx was, studying and writing away in the British Library. In fact, we would be wise to look in new places—in Asia or Africa—for the next Marx, Marie Curie, or Jefferson. The secret to distinguishing them and their ideas is, of course, that which is the secret to coming up with those ideas in the first place. It is what my father devoted his life to as a scientist. It is what sets apart all creative thinkers: it is asking the right questions. It is looking at a catastrophe like the Black Plague and asking how our world has changed. How do we cope with the new reality? How do we reorganize? What must we do differently? Because with the right questions come breakthroughs and transformational eras, like the Renaissance. And only by asking important questions during these transformational eras can we hope for real progress and avoid calamity.

That is why I am certain that, were he alive today and looking at the information revolution in which he was involved throughout his life, my father would be asking questions like those posed in these pages and forcing our perspectives away from the daily headlines and press releases of the self-interested toward more fundamental issues. What is really changing? Why is it important? What do we want?

As the years have passed, I have come to appreciate my father in many ways that eluded me when I was younger. As a scientist, he tried to look at the world—whether in the laboratory or in a history book—with a kind of detachment and skepticism of conventional explanations. History is spin, after all. It is tailored to the needs of its authors, those who commissioned them, and the times. Progress is sometimes not the result of enlightened leadership, inspired plans, or battlefield victories. Sometimes—perhaps more often than we care to acknowledge—it comes from accident, chance, or even catastrophe.

Part of my father's message, therefore, was cautionary. It was, if anything, a warning we need to heed today more than at any other time in memory. That is because the press of great technological, demographic, and natural changes has us on the verge of a watershed moment in the long story of civilization.

We should resist the temptation to judge history in the moment; when we are too caught up in the human reaction to events as they occur, we lack the objectivity to see what progress they might trigger. The good is often unleashed by the bad, and big changes come from small or unexpected developments—though any meteorologist will tell you that 85 percent of the time tomorrow we will have the same weather as today, that means that one out every six or so days we do not.

Most of the time, the changes that take place impact us only slightly. History unfolds incrementally. Decades go by in which it seems very little happens. This leads to the most common heuristic trap (the most common shortcut analysts take) when trying to predict what will happen in the future: we assume that it will be much like yesterday. You know the phenomenon. It manifests itself in our daily life in countless ways. For example, a guy tries to blow up an airplane with a bomb in his shoe, and we therefore determine that the big new threat of our time is shoe bombers. Thereafter, you must take your shoes off every time you pass through airport security.

Past results are not a predictor of future performance. In fact, we are deceived by the past into thinking in the wrong ways about what is yet to come. We call it *experience*. We can only project the future based on our experience. It is a useful survival instinct. It keeps us from stumbling upon the same bear in the same part of the woods whenever we go for a walk. But it doesn't prepare us for what comes when we leave the woods with which we are familiar and go to a place where there are no bears and where the threats and opportunities are entirely new to us.

That does not mean that we must be blindsided by the future. If we train ourselves to think with the distance and skepticism and sensitive eye of the scientist, we can often see patterns in the past that can help us navigate—and even predict and prepare for—very different futures. No one has, of course, perfected this skill. But those who are better at it end up with a big advantage, and are thus better able to thrive as change inevitably comes—even very, very big change.

There are many brilliant minds in the world today asking just such questions. But it will take many more at every level of society to tap the potential this new era may bring. Be among them. Remember Einstein's admonition: when searching for solutions, devote your attention to finding the right question first. If you do, the best answers are sure to follow.

ACKNOWLEDGMENTS

I once worked for a man who regularly liked to append to his overly long letters the old line that if he had had more time he would have written a shorter letter. Writing this book has demonstrated to me that what's true for letters is certainly true for books. Writing this short one took more time in many ways than some of the longer ones I have done . . . and for good reason. Each word counts for more and keeping the structure simple requires discipline.

Fortunately for me, I had the support and guidance of the great team at TED Books, led by Michelle Quint, who is as good an editor as I have ever had the privilege of working with. That means she is smart, patient, and able to maintain her sense of humor in all situations. I'm grateful to her and to everyone at TED, including, notably, Chris Anderson, who first asked me to give the TED Talk that led to this book, and who has taught me a great deal about how a great communications organization should be run.

My agent, Esmond Harmsworth, was, as ever, a vital support throughout the process, as have been my colleagues at Foreign Policy. From the leadership team at Graham Holdings, our parent company, including Don Graham and Tim O'Shaughnessy, to each and every member of the FP team, I get more intellectual stimulation and have more good fun working with them than anyone deserves in a job they also get paid for. My assistants during the production of this book, including especially Cathryn Hunt, were enormously supportive and would certainly deserve the first Nobel Prize for assistanting as soon as that is created.

I also want to thank the Carnegie Endowment for International Peace, where I have long been a visiting fellow for support and to acknowledge the great leadership there of ambassador William Burns. My program there is supported by my good friend, the great philanthropist and thinker Bernard L. Schwartz, and I am deeply grateful to him not just for the support but for the many long conversations we have had on the subjects covered herein. Finally, I would also like to express my thanks to my colleagues at Columbia University's School of International and Public Affairs, notably Dean Merit Janow.

Others who have helped me hone these ideas in conversation are too numerous to mention here, but trust me, I will thank each of you in person. Two ongoing friendships have played an especially important role in shaping my views on the subjects in this book, and those are the ones I enjoy with Tom Friedman and my longtime colleague and mentor, Jeff Garten. I will be forever grateful to them for helping me to grow intellectually and for regularly making me look much smarter than I am.

Naturally, the most patient, kind, loving, and wise people I know are my family, and for them there is more than thanks—there is also love. My mom, a writer, inspired me to do this for a living. My brother and sister have always been there for me. My daughters, Joanna and Laura, are my pride and joy, the reasons I think about the future and feel it is so vital we fight and scrape and innovate to make the best one we can. They are smart and talented and beautiful and kind and more reward than any one person deserves in a lifetime. And then, compounding that great good fortune, there is Carla. She not only fills me with love but she inspires me in infinite ways. To live and travel and enjoy life with a partner who is also a muse and wise advisor and the world's greatest expert on where to find the best snacks . . . that's more than I ever dreamt possible. To her, to Joanna and Laura, and to everyone else noted here or unjustly overlooked, you have my deepest gratitude.

ABOUT THE AUTHOR

David J. Rothkopf is CEO and editor of the FP Group, which publishes *Foreign Policy* magazine and the foreignpolicy.com website, presents FP Events worldwide, and produces research through FP Analytics. He is also the host of *The Editor's Roundtable* (*The E.R.*) podcast. David is a widely published author and has written numerous books, including most recently *National Insecurity: American Leadership in an Age of Fear*, as well as *Power, Inc.*, *Superclass*, and *Running the World*. He is a visiting professor of international and public affairs at Columbia University's School of International and Public Affairs and a visiting scholar at the Carnegie Endowment for International Peace.

WATCH DAVID ROTHKOPF'S TED TALK

David's TED Talk, available for free at TED.com, is the companion to *The Great Questions of Tomorrow*.

PHOTO: BRET HARTMAN/TED

RELATED TALKS

Gordon Brown
Wiring a Web for Global Good
We're at a unique moment in history,
says former UK prime minister
Gordon Brown: we can use today's
interconnectedness to develop our
shared global ethic and work together
to confront the challenges of poverty,
security, climate change, and the
economy.

Neha Narula
The Future of Money
What happens when the way we buy,
sell, and pay for things changes,
perhaps even removing the need for
banks or currency-exchange bureaus?
That's the radical promise of a world
powered by cryptocurrencies like
Bitcoin and Ethereum. We're not
there yet, but in this sparky talk, digital currency researcher Neha Narula
describes the collective fiction of
money — and paints a picture of a very
different-looking future.

Paddy Ashdown
The Global Power Shift
Paddy Ashdown believes we are living
in a moment in history when power is
changing in ways it never has before.
In a spellbinding talk, he outlines the
three major global shifts that he sees
coming.

David Cameron
The Next Age of Government
The former UK prime minister says
we're entering a new era in which
governments themselves have less
power (and less money) and people
empowered by technology have more.
Tapping into new ideas on behavioral
economics, David Cameron explores
how these trends could be turned into
smarter policy.

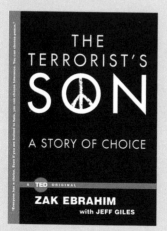

The Terrorist's Son:
A Story of Choice
by Zak Ebrahim

The astonishing first-person
account of an American boy
raised on dogma and hate—a boy
presumed to follow in his father's
footsteps—and the man who
chose a different path. Ebrahim
argues that everyone has a choice.
Even if you're raised to hate, you
can choose tolerance. You can
choose empathy.

Payoff:
The Hidden Logic That
Shapes Our Motivations
by Dan Ariely

Payoff investigates the true
nature of motivation, our partial
blindness to the way it works, and
how we can bridge this gap. Dan
Ariely digs deep to find the root
of motivation—how it works and
how we can use this knowledge to
approach important choices in our
own lives.

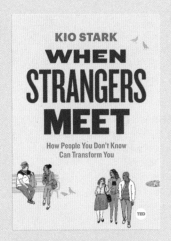

***When Strangers Meet:
How People You Don't Know
Can Transform You***
by Kio Stark

When Strangers Meet reveals the transformative possibility of talking to people you don't know—how these beautiful interruptions in daily life can change you and the world we share. Kio Stark argues for the surprising pleasures of talking to strangers.

Asteroid Hunters
by Carrie Nugent

Everyone's got questions about asteroids. What are they, and where do they come from? And most urgently, are they going to hit earth? Asteroid hunter Carrie Nugent reveals everything we know about asteroids, and how new technology may help us prevent a natural disaster.

ABOUT TED BOOKS

TED Books are small books about big ideas. They're short enough to read in a single sitting, but long enough to delve deeply into a topic. The wide-ranging series covers everything from architecture to business, space travel to love, and is perfect for anyone with a curious mind and an expansive love of learning.

Each TED Book is paired with a related TED Talk, available online at TED.com. The books pick up where the talks leave off. An eighteen-minute speech can plant a seed or spark the imagination, but many talks create a need to go deeper, to learn more, to tell a longer story. TED Books fill this need.

ABOUT TED

TED is a nonprofit organization devoted to spreading ideas, usually in the form of short, powerful talks (of eighteen minutes or less), but also through books, animation, radio programs, and events. TED began in 1984 as a conference where Technology, Entertainment, and Design converged, and today covers almost every topic—from science to business to global issues—in more than one hundred languages. Meanwhile, independently run TEDx events help share ideas in communities around the world.

TED is a global community, welcoming people from every discipline and culture who seek a deeper understanding of the world. We believe passionately in the power of ideas to change attitudes, lives, and, ultimately, our future. On TED.com, we're building a clearinghouse of free knowledge from the world's most inspired thinkers—and a community of curious souls who engage with ideas and one another, both online and at TED and TEDx events around the world, all year long.

In fact, everything we do—from the TED Radio Hour to the projects sparked by the TED Prize, from the global TEDx community to the TED-Ed lesson series—is driven by this goal: how can we best spread great ideas?

TED is owned by a nonprofit, nonpartisan foundation.